Catastrophes!

Catastrophes!

Earthquakes, Tsunamis, Tornadoes, and Other Earth-Shattering Disasters

Donald R. Prothero

With Illustrations by

Pat Linse

THE JOHNS HOPKINS UNIVERSITY PRESS

Baltimore

© 2011 The Johns Hopkins University Press
All rights reserved. Published 2011
Printed in the United States of America on acid-free paper
2 4 6 8 9 7 5 3 1

The Johns Hopkins University Press
2715 North Charles Street
Baltimore, Maryland 21218-4363
www.press.jhu.edu

Library of Congress Cataloging-in-Publication Data
Prothero, Donald R.
Catastrophes! : earthquakes, tsunamis, tornadoes, and other earth-shattering disasters /
Donald R. Prothero.
p.   cm.
Includes bibliographical references and index.
ISBN-13: 978-0-8018-9692-7 (hardcover : alk. paper)
ISBN-10: 0-8018-9692-4 (hardcover : alk. paper)
1. Catastrophes (Geology) I. Title.
QE506.P76 2011
551—dc22   2010019631

A catalog record for this book is available from the British Library.

*Special discounts are available for bulk purchases of this book. For more information,
please contact Special Sales at 410-516-6936 or specialsales@press.jhu.edu.*

The Johns Hopkins University Press uses environmentally friendly book materials,
including recycled text paper that is composed of at least 30 percent post-consumer
waste, whenever possible. All of our book papers are acid-free, and our jackets and
covers are printed on paper with recycled content.

To my amazing wife,
Dr. Teresa LeVelle,
for all her encouragement and inspiration

Civilization exists by geologic consent,
subject to change without notice.

—Will Durant, historian

Contents

Preface

The idea for this book originated from the horrors of the December 26, 2004, Indian Ocean tsunami. My wife and I watched the news coverage and thought that there was a need for such a book. This need was further emphasized by the 2005 Hurricane Katrina disaster, the 2008 Mississippi Valley floods, and the 2010 Haiti earthquake. However, my heavy teaching load and many other research and writing commitments prevented me from pursuing this project until a summer without a field season (2009) allowed me time to work on it. I've been teaching about these events in my introductory physical geology courses for more than 25 years now, but this book gave me a chance to expand on the usual material and to take a novel, more paleontological approach to natural disasters.

I thank my editors at the Johns Hopkins University Press, especially editor Vincent J. Burke and copyeditor Andre Barnett, for all their help with producing this book. I thank Eddie Bromhead, Chuck Chapin, and two anonymous reviewers for their comments on the earlier drafts. I thank Kristine Wendt for her wonderful artwork. I especially thank Pat Linse for her incredible transformation of the art in the book, and my wife for cleaning up and improving many of the images.

Finally, I am grateful to my wonderful family for letting me focus on this project in peace and quiet while I wrote it over a few weeks in July 2009. It is always fun to show my incredible sons Erik, Zachary, and Gabriel the power of the earth and wonders of rocks and fossils. This work would not have been possible without the loving support of my wife, Dr. Teresa LeVelle, who did everything she could to help me concentrate and to keep at my writing when many other distractions beckoned.

Catastrophes!

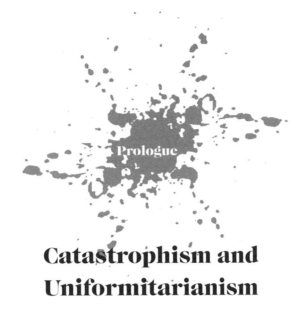

Catastrophism and Uniformitarianism

In the mid-1700s, western European thought was at a crossroads. For more than a thousand years, all scholarship and learning in Europe was the domain of priests and clerics, because they were practically the only people with at least some education. Most people in the world were illiterate, not able to read even their native language. Trained by their church, these clerics framed all thought in terms of the Bible, and the Bible was their first source of truth, whether for historical events or as an explanation of nature. Almost all scholarship focused on pointless theological debates (the well-known example: "How many angels could dance on the head of a pin?"), and scholars slavishly copied and believed every ancient dogma that the church approved. Medieval scholars were especially fond of Aristotle and took his natural history literally, even as the evidence from real organisms showed that it was manifestly false.

Biblical stories were taken at face value, and anyone who dared to dissent from this viewpoint could suffer severely. Many, such as philosopher and astronomer Giordano Bruno and others, were burned at the stake for heresy. Galileo spent the last years of his life under house arrest for suggesting that the earth was not the center of the universe. Copernicus wisely did not release his ideas for publication until he was dying, because he rightly feared that he would be tortured to death if they were published while he was alive. There were a few Renaissance humanist scholars, such as Petrarch, who coined the term *dark ages* for the millennium of medieval ignorance after the fall of the Roman Empire, but even they were usually trained as priests and did not challenge Catholic Church dogma.

In the 1600s and early 1700s, the works of giants such as Isaac Newton, Francis Bacon, and René Descartes had shown the power of explaining nature by mathematical methods and natural law. Even though he was a devout man, Newton did not require God to move the planets and stars—for him, the natural laws of gravity were a simpler, more natural and better explanation. Natural history was changing, with the excitement of discovery and the description of animals and plants that did not conform to Aristotle's views. Philosophers such as Immanuel Kant, Baruch de Spinoza, Descartes, Gottfried Wilhelm Leibniz, and Johann Wolfgang von Goethe boldly speculated on matters such as free will and determinism, the nature of good and evil, and how to live a moral life, often resorting entirely to logic without invoking the Bible or God. Descartes' famous maxim, *Cogito ergo sum* (I think, therefore I am), is a central statement of this philosophy. Political theorists such as John Locke in England and Montesquieu in France began to talk about human rights, natural liberty, separation of powers, freedom of speech and thought, and they openly questioned the hierarchy of power of kings, noblemen, and the church over the common people.

The mid- and late 1700s saw a movement throughout Western Europe known as "the Enlightenment." The term has been defined in many ways, but the basic idea was that human logic and reason were the primary method of finding the truth about the world, not adherence to ancient documents like the Bible or the writings of Aristotle. New scholars built on the earlier thoughts of philosophers of the late 1600s and early 1700s, and these ideas of reason and natural law advanced in nearly every European center of learning. France saw the iconoclastic writings of Voltaire, Rousseau, Diderot, Buffon, as well as questioning by many others of the power of the Catholic Church and the King, arguing for a world based on reason and humanism without the powerful political and philosophical stranglehold of the church. England has its share of Enlightenment thinkers, especially in the political sphere, and they in turn influenced Benjamin Franklin, Thomas Jefferson, and Thomas Paine, who used these principles to establish a moral argument for the American Revolution and designed a new form of government enshrined in the American Constitution.

Many other countries saw the effects of the Enlightenment scholars, but a particularly interesting example was the Scottish Enlightenment. In the mid- and late 1700s, Scotland was truly a remarkable hotbed of intellectual ferment. Edinburgh and Glasgow boasted some of the best universities in the world, and Scotland had one of the highest literacy rates in the world (about 75% in 1750). Legends of scholarship all lived in Edinburgh or Glasgow, and these men met at formal social clubs, such as the Select Society and the Poker Club, where they often exchanged radical ideas. Some of these famous minds included the poet Robert Burns, the skeptical philosopher David Hume, the founder of modern capitalism Adam Smith, the inventor of the steam engine James Watt, the founder of thermochemistry Joseph Black, and the philosopher Francis Hutcheson.

Among these Scottish thinkers was a doctor and landowner named James Hutton. Originally trained in a broad range of Enlightenment subjects, including law, theology, and chemistry, he spent most of his life managing the farms he'd inherited, which gave him a comfortable income. He was particularly close to chemist Joseph Black, and the two of them socialized with economist Adam Smith and eventually founded the Oyster Club for their weekly meetings. By 1753, he became interested in the study of the earth's surface and constantly surveyed every exposure (whether pit, ditch, or riverbank) to see what it revealed. He invested heavily in canal building and used his growing understanding of the earth to help the canal engineers make better decisions.

As his ideas developed, he began to apply the Enlightenment emphasis on natural law to geology. He saw the earth as produced by natural processes like those operating today, not supernaturally created in only 6,000 years as the church maintained. He was no atheist but was devoutly religious. Nonetheless, he realized that science cannot make progress if it resorts to supernatural dogma or just says, "God did it" and stops looking. The Scientific Revolution was based on the principle of *methodological naturalism*. As scientists, we must work on scientific problems assuming that natural law applies, and we are not allowed to use supernatural explanations as an easy way out. Once you introduce a supernatural explanation to a scientific discussion, there is no way to test the hypothesis by scientific methods. A famous Sidney Harris cartoon says it best. It shows two scientists at a blackboard, pondering a complex set of equations, with the phrase "Then a miracle occurs" right in the middle of the equation. In the caption, one scientist tells the other, "I think you should be more explicit here in step two." Applying this naturalistic assumption to the rocks around Scotland and the rest of the British Isles, he saw the evidence for a long, slow process of earth change and development. Hutton visited Hadrian's Wall, which the Romans had built to keep the marauding Scotsmen out of their English domain, and realized that it showed almost no signs of change over the nearly 2,000 years after it had been erected. Studying the sand grains accumulating in a river or beach, Hutton reasoned that it must have taken thousands to millions of years for thick sheets of sandstone to form by such slow processes. Looking at tilted rock layers, he reasoned that they must have once formed as horizontal beds of sand and mud on the seafloor, hardened into sandstone and mudstone, and then slowly tilted on edge by mountain building. In many cases, these tilted layers were eroded off, and new layers deposited on top of them, which required even more time. Hutton looked at basaltic lava flows (which most scholars thought were deposited out of the waters of Noah's Flood) and realized that they had once been molten lava and had intruded into the surrounding rock and baked the edges of the surrounding rock with their immense heat.

Little by little, Hutton's naturalistic observations developed into what we now call *uniformitarianism*, or the uniform application of the laws of nature across space and time. Uniformitarianism is often summarized as "the present is the key to the past" because scientists must use present-day laws and processes to understand past events. Hutton's observations led him to realize that the earth was immensely old, with "no vestige of a beginning, no prospect of an end." This was in direct conflict with the religious authorities of the day, who still asserted that the earth was created supernaturally in 4004 BC. It was also in conflict with many of his contemporaries in geology, such as Abraham Gottlob Werner of the Freiburg Mining Academy, who asserted that lava flows came out of a body of water. Consequently, the uniformitarian position was considered the antithesis of *catastrophism*: that supernatural catastrophes like Noah's Flood explained the earth's history.

Uniformitarianism is not restricted to geology, however. Anytime science works on things we cannot observe directly, we must assume uniformitarianism and that natural law applies. For example, most of astronomy concerns bodies that are thousands to millions of light-years away and cannot be experimented on directly, because the light we are receiving from them was emitted thousands to millions of years ago and is just now reaching us. For most of the past two centuries, everything we knew about the behavior of atoms and molecules had to be inferred indirectly by experiment, because we could not observe a chemical reaction or the structure of an atom directly. Much of what we know in biology is also inferred using the uniformitarian assumption, because it is too tiny, too slow, or too subtle for us to observe directly. Indeed, nearly all of science (except for a handful of experiments in physics and biology that can be run in real time) requires this assumption.

Hutton published his ideas in a 1788 essay, "Theory of the Earth, or an Investigation of the Laws observable in the Composition, Dissolution, and Restoration of Land upon the Globe." He then wrote about it at greater length in his 1795 book *Theory of the Earth with Proof and Illustrations*. Unfortunately, his prose was difficult to read, and his ideas were pretty challenging and revolutionary, so they did not immediately convince the scientific community. In 1802, his student John Playfair published *Illustrations of the Huttonian Theory of the Earth,* which helped clarify and argue Hutton's concepts more convincingly. The final step to modern geology, however, did not occur until a generation later, when a young barrister named Charles Lyell abandoned his training in law and became an enthusiastic geologist. By 1827, he was spending his time and resources on geological excursions all over Europe (often accompanied by other pioneering geologists, such as Roderick Impey Murchison). From this research came the landmark book *Principles of Geology, being an attempt to explain the former changes of the Earth's surface, by reference to causes now in operation,* published in three volumes from 1830 to 1833. As the subtitle implies, Lyell wanted to carry Hutton's uniformitarian principles to their complete and logical conclusion and forever exclude supernatural causes from geology. Trained as a lawyer, Lyell was expert in arguing a case clearly and logically and in shading his argument to discredit the ideas of his opponent. Indeed, *Principles of Geology* is virtually a legal brief on a scientific topic, argued in a one-sided fashion to make the opponents' position seem ridiculous and illogical. (As most lawyers know, legal "briefs" are anything but brief, but often run to book length.)

Lyell's argument was so successful and skillfully skewered the last remnants of supernaturalism that geology never looked back. He not only convinced everyone in the geological community of the time but also influenced many others. One of these was the young Charles Darwin, who read Lyell's books on his long voyage on the HMS *Beagle* around the world from 1831 to 1836. From these books, he gained a uniformitarian perspective on geology and biology by looking "through Lyell's eyes" at the amazing sights he witnessed. These Lyellian ideas led Darwin not only to explain correctly the coral atolls of the Pacific by uniformitarian means but also to see all of life as controlled by natural law, which eventually led to his discovery of natural selection and to the publication of his ideas on evolution. Lyell, by the way, became a close friend and mentor of Darwin well before the publication of *On the Origin of Species* in 1859 and supported his rise through the scientific ranks. Lyell actually helped mediate the complex issues raised by Alfred Russel Wallace's independent discovery of the same idea in 1858.

In *Principles of Geology* (1830–1833), Lyell worked hard to show not only that the uniformity of natural causes was sufficient to explain all geologic phenomena (*actualism*) but also that there was no reason to believe that these processes had ever operated at extraordinarily high rates or scales (*gradualism*).

In his battle to discredit old-fashioned biblical supernatural catastrophes such as Noah's Flood, Lyell deliberately conflated these two different concepts of applying natural law to geology. Consequently, geologists for generations believed that large-scale events were simply impossible because they violated Lyell's gradualistic uniformitarianism.

In retrospect, we now know that gradualism is not required because gigantic events such as meteorite impacts are rare but do happen and do not violate natural law. All of the catastrophes described in this book are strictly natural and require only application of natural law to understand them. Many of them operate on huge scales that have never been witnessed in human history, but they are still naturally occurring. In some cases (see chapter 5), the geological community mistakenly refused to recognize huge natural events such as the Scablands Floods because they were misled by Lyell's mistaken argument that all uniformitarian events also had to be gradual. As we shall see in chapter 11, the "new catastrophism" that followed the discovery of the impact event at the end of the age of dinosaurs led many people to overemphasize the extreme and unusual events (like impacts) in geologic history, although that fad has apparently died down.

As a coda, we must not confuse these natural shifts in emphasis about gradualism and natural catastrophes of geology with the bizarre ideas of American fundamentalists and their strange creation known as "flood geology." It is common for creationists to pull quotes out of context about the gradualism-catastrophism debate to attempt to support their ideas about supernatural events, but nothing could be further from the truth. Supernatural events such as Noah's Flood were thoroughly debunked and discredited by good Christian men in the early 1800s and completely abandoned as a reasonable description of the earth by the publication of Lyell's book in 1830–1833. Even though many devout geologists like William Buckland, Murchison, and Adam Sedgwick were still good Christians, they knew that the record of earth history did *not* support the idea of Noah's Flood.

Ironically, these creationists dishonestly claimed that the sequence of fossils through time is a fabrication cooked up by evolutionists to prove their case. If they knew the tiniest bit of history, they would realize that this is patently false, because the sequence of fossils was a demonstrable fact of nature first shown by devout men such as William Smith in the 1790s, a full half-century before Darwin's book. These modern anti-scientists make many other absurd claims about geology and the rock record that are thoroughly debunked in chapter 3 of my 2007 book, *Evolution: What the Fossils Say and Why It Matters*, so I will not repeat that entire discussion here. Suffice it to say that if devout Christian geologists of the 1830s and 1840s such as Buckland, Lyell, or Sedgwick saw what these pseudoscientists (who never look at real outcrops and don't know the first thing about geology) have written, they would be rolling over in their graves!

Earthquakes

The Earth in Upheaval

A bad earthquake at once destroys our oldest associations: the earth, the very
emblem of solidity, has moved beneath our feet like a thin crust over fluid;
one second of time has created in the mind a strong idea of insecurity,
which hours of reflection would never have produced.

—Charles Darwin, *The Voyage of the Beagle*, 1839

The End of Optimism: Lisbon, 1755

November 1, 1755, was All Saints' Day, one of the holiest days on the Roman
Catholic calendar. In the mighty city of Lisbon, Portugal, the faithful were
crowding the streets. They were heading toward the great cathedrals and
smaller churches that had been built over 200 years by the wealth of the Portuguese trade empire that ranged from Brazil to southern Africa and from
India to China. Lisbon was one of the largest and richest cities in Europe, with
more than a quarter of a million people and many magnificent palaces and
public buildings. However, the Portuguese empire had begun to rot and stag-
nate since its prime two centuries earlier. Downtown was a medieval warren

of narrow, crooked streets, penetrating poorly built slums, with no sewage system or reliable fresh water. Lisbon's poor were exposed to disease and misery, compounded by a social system ruled by an ineffective monarchy and effete noblemen competing for the biggest palaces and fanciest balls while riding in coaches or sedan chairs to avoid the filthy streets. The educational system was practically nonexistent; therefore, most people were ignorant and superstitious. Portugal had not fostered a middle class of merchants and artisans to bridge the gap between rich and poor, relegating those trading jobs to foreign merchants. The Catholic Church, and especially the Jesuits, held absolute power over the people and the educational system. The church enforced its power with the bloody Inquisition, routinely torturing Jews and other heretics by breaking them on the rack and burning people at the stake for the slightest transgression from orthodoxy.

This was the situation on November 1, 1755. By 9:30 a.m., most of the population was already packed into the church pews or in the streets on their way to mass. The day had dawned crisp and cloudless. The priests had just begun chanting "*Gaudeamus omnes in Domino, diem festum . . .*" when the cathedrals and churches started to sway and lurch from side to side, like ships tossed on the waves (fig. 1.1). The bronze bells in the towers began to ring madly, and the many candles in the churches toppled over. Huge blocks of masonry fell, crushing fearful worshipers as they prayed. Those who rushed outside found the city enveloped in a huge dust cloud, making the morning almost as dark as midnight. Many died as the second larger shock struck a few minutes later, toppling most of the buildings that had managed to withstand the first tremor.

Those who could run from the collapsing city and dust clouds sought refuge along the quays of the harbor. However, about 90 minutes after the first great shock, the first of three huge tsunamis (seismic sea waves) rolled in and washed nearly everyone in the port out to sea. These waves sank all but the largest merchant ships and warships and washed away the docks and warehouses and nearly all the coastal infrastructure. Meanwhile, back in the city, the many fallen candles and broken hearths set fire to the highly flammable wooden buildings, and soon the entire city was engulfed in a conflagration

Fig. 1.1. Artist's conception of the 1755 Lisbon earthquake. (Copper engraving, France; original in Museu da Cidade, Lisbon. Reproduced in O Terramoto de 1755; Testamunhos Britanicos. Courtesy Pacific Earthquake Engineering Research Center Earthquake Engineering Online Archive.)

that destroyed everything combustible, leaving only the stone ruins unburned. Many of the victims trapped in the rubble were asphyxiated or burned to death by the ensuing fires. There was no one to rescue them, and the handful of able-bodied survivors were too stunned and frightened to mount much of a rescue effort or attempt to fight the flames when the city's water system had been destroyed.

This eyewitness account by Charles Davy, a British merchant living in Lisbon, is particularly horrifying:

There never was a finer morning seen than the 1st of November; the sun shone out in its full luster; the whole face of the sky was perfectly serene and clear; and not the least signal of warning of that approaching event, which has made this once flourishing, opulent, and populous city, a scene of the utmost horror and desolation, except only such as served to alarm, but scarcely left a moment's time to fly from the general destruction.

It was on the morning of this fatal day, between the hours of nine and ten, that I was set down in my apartment, just finishing a letter, when the papers and table I was writing on began to tremble with a gentle motion, which rather surprised me, as I could not perceive a breath of wind stirring. Whilst I was reflecting with myself what this could be owing to, . . . the whole house began to shake from the very foundation, which at first I imputed to the rattling of several coaches in the main street, which usually passed that way, at this time, from Belem to the palace; but on hearkening more attentively, I was soon undeceived, as I found it was owing to a strange frightful kind of noise under ground, resembling the hollow distant rumbling of thunder. All this passed in less than a minute, and I must confess I now began to be alarmed, as it naturally occurred to me that this noise might possibly be the forerunner of an earthquake . . .

. . . The house I was in shook with such violence, that the upper stories immediately fell; and though my apartment (which was the first floor) did not then share the same fate, yet everything was thrown out of its place in such a manner that it was with no small difficulty I kept my feet, and expected nothing less than to be soon crushed to death, as the walls continued rocking to and fro in the frightfulest manner, opening in several places; large stones falling down on every side from the cracks, and the ends of most of the rafters starting out from the roof. To add to this terrifying scene, the sky in a moment became so gloomy that I could now distinguish no particular object . . . I hastened out of the house and through the narrow streets, where the buildings either were down or were continually falling, and climbed over the ruins of St. Paul's Church to get to the river's side, where I thought I might find safety. Here I found a prodigious concourse of people of both sexes, and of all ranks and conditions, among whom I observed some of the principal canons of the patriarchal church, in their purple robes and rochets, as these all go in the habit of bishops; several priests who had run from the altars in their sacerdotal vestments in the midst of their celebrating Mass; ladies half dressed, and some without shoes; all these, whom their mutual dangers had here assembled as to a place of safety, were on their knees at prayers, with the terrors of death in their countenances, every one striking his breast and crying out incessantly, *Misericordia meu Dios!* . . . In the midst of our devotions, the second great shock came on, little less violent than the first, and completed the ruin of those buildings which had been already much shattered. The consternation now became so universal that the shrieks and cries of Miserecordia could be distinctly heard from the top of St. Catherine's Hill, at a considerable distance off, whither a vast number of people had likewise retreated; at the same time we could hear the fall of the parish church there, whereby many persons were killed on the spot, and others mortally wounded.

You may judge of the force of this shock, when I inform you it was so violent that I could scarce keep on my knees; but it was attended with some circumstances still more dreadful than the former. On a sudden I heard a general outcry, "The sea is coming in, we shall be all lost." Upon this, turning my eyes towards the river, which in that place is nearly four miles broad, I could perceive it heaving and swelling in the most unaccountable manner, as no wind was stirring. In an instant there appeared, at some small distance, a large body of water, rising as it were like a mountain. It came on foaming and roaring, and rushed towards the shore with such impetuosity, that we all immediately ran for our lives as fast as possible; many were actually swept away, and the rest above their waist in water at a good distance from the banks. For my own part I had the narrowest escape, and should certainly have been lost, had I not grasped a large beam that lay on the ground, till the water returned to its channel, which it did almost at the same instant, with equal rapidity. As there now appeared at least as much danger from the sea as the land, and I scarce knew whither to retire for shelter, I took a sudden resolution of returning back, with my clothes all dripping, to the area of St. Paul's. Here I stood some time, and observed the ships tumbling and tossing about as in a violent storm; some had broken their cables, and were carried to the other side of the Tagus; others were whirled around with incredible swiftness; several large boats were turned keel upwards; and all this without any wind, which seemed the more astonishing. It was at the time of which I am now speaking, that the fine new quay, built entirely of rough marble, at an immense expense, was entirely swallowed up, with all the people on it, who had fled thither for safety, and had reason to think themselves out of danger in such a place: at the same time, a great number of boats and small vessels, anchored near it (all likewise full of people, who had retired thither for the same purpose), were all swallowed up, as in a whirlpool, and nevermore appeared.

This last dreadful incident I did not see with my own eyes, as it passed three or four stones' throws from the spot where I then was; but I had the account as here given from several masters of ships, who were anchored within two or three hundred yards of the quay, and saw the whole catastrophe. One of them in particular informed me that when the second shock came on, he could perceive the whole city waving backwards and forwards, like the sea when the wind first begins to rise; that the agitation of the earth was so great even under the river, that it threw up his large anchor from the mooring, which swam, as he termed it, on the surface of the water: that immediately upon this extraordinary concussion, the river rose at once near twenty feet, and in a moment subsided; at which instant he saw the quay, with the whole concourse of people upon it, sink down, and at the same time every one of the boats and vessels that were near it was drawn into the cavity, which he supposed instantly closed upon them, inasmuch as not the least sign of a wreck was ever seen afterwards . . .

I had not been long in the area of St. Paul's when I felt the third shock, somewhat less violent than the two former, after which the sea rushed in again, and retired with the same rapidity, and I remained up to my knees in water, though I had gotten upon a small eminence at some distance from the river, with the ruins of several intervening houses to break its force. At this time I took notice the waters retired so impetuously, that some vessels were left quite dry, which rode in seven fathom water; the river thus continued alternately rushing on and retiring several times together, in such sort that it was justly dreaded Lisbon would now meet the same fate which a few years before had befallen the city of Lima; and no doubt had this place lain open to the sea, and the force of the waves not been somewhat broken by the winding of the bay, the lower parts of it at least would have been totally destroyed . . .

. . . There was another great shock after this, which pretty much affected the river, but I think not so violently as the preceding; though several persons assured me that as they were riding on horseback in the great road leading to Belem, one side of which lies open to the river, the waves rushed in with so much rapidity that they were obliged to gallop as fast as possible to the upper grounds, for fear of being carried away.

I was now in such a situation that I knew not which way to turn myself: if I remained there, I was in danger from the sea; if I retired farther from the shore, the houses threatened certain destruction; and at last, I resolved to go to the Mint, which being a low and very strong building, had received no considerable damage, except in some of the apartments towards the river. The party of soldiers, which is every day set there on guard, had all deserted the place, and the only person that remained was the commanding officer, a nobleman's son . . . We both retired inward to a hillock of stones and rubbish . . . In short, it was owing to the magnanimity of this young man that the Mint, which at this time had upwards of two millions of money in it, was not robbed; and indeed I do him no more than justice in saying that I never saw any one behave with equal serenity and composure on occasions much less dreadful than the present . . .

16 Catastrophes!

Perhaps you may think the present doleful subject here concluded; but alas! the horrors of the 1st of November are sufficient to fill a volume. As soon as it grew dark, another scene presented itself little less shocking than those already described: the whole city appeared in a blaze, which was so bright that I could easily see to read by it. It may be said without exaggeration, it was on fire at least in a hundred different places at once, and thus continued burning for six days together, without intermission, or the least attempt being made to stop its progress.

It went on consuming everything the earthquake had spared, and the people were so dejected and terrified that few or none had courage enough to venture down to save any part of their substance; every one had his eyes turned towards the flames, and stood looking on with silent grief, which was only interrupted by the cries and shrieks of women and children calling on the saints and angels for succor, whenever the earth began to tremble, which was so often this night, and indeed I may say ever since, that the tremors, more or less, did not cease for a quarter of an hour together . . .

But what would appear incredible to you, were the fact less public and notorious, is that a gang of hardened villains, who had been confined and got out of prison when the wall fell, at the first shock, were busily employed in setting fire to those buildings which stood some chance of escaping the general destruction. I cannot conceive what could have induced them to this hellish work, except to add to the horror and confusion that they might, by this means, have the better opportunity of plundering with security . . . The fire, in short, by some means or other, may be said to have destroyed the whole city, at least everything that was grand or valuable in it.

With regard to the buildings, it was observed that the solidest in general fell the first. Every parish church, convent, nunnery, palace, and public edifice, with an infinite number of private houses, were either thrown down or so miserably shattered that it was rendered dangerous to pass by them.

The whole number of persons that perished, including those who were burnt or afterwards crushed to death whilst digging in the ruins, is supposed, on the lowest calculation, to amount to more than sixty thousand; and though the damage in other respects cannot be computed, yet you may form some idea of it when I assure you that this extensive and opulent city is now nothing but a vast heap of ruins; that the rich and the poor are at present upon a level; some thousands of families which but the day before had been easy in their circumstances, being now scattered about in the fields, wanting every conveniency of life, and finding none able to relieve them. (Tappan 1914, 5:618–628)

Within a few days, the fires had burned out, and the survivors were mostly huddled on the outskirts of the city, suffering from shock and various injuries. Most were praying for their lives, convinced that the great catastrophe was a punishment for their sins. By a stroke of good luck, King José I and the royal family were out of the city in their royal retreat in Belém and were able to run to the safety of their palace gardens as it too collapsed. The Jesuit priests in the king's court and in the city were all commanding the citizens to do nothing to help themselves but to continue to pray and do penance for their sins. In their view, the earthquake was a punishment, and any attempt to circumvent God's will was sacrilegious. Fortunately for Portugal, the king's secretary of state was Sebastiao José Carvalho é Melo, a well-educated and practical man who had seen much of London, Vienna, and the more advanced cities of Europe. He had long chafed at the superstition and backwardness of Portugal. He immediately calmed the panicked and paralyzed king, and when asked what to do, said, "Bury the dead and feed the living." The king gave him complete authority to act in his royal name. Carvalho lost no time in organizing rescue parties, rounding up the military to establish martial law and shoot the looters and criminals roaming the streets, and imposing order. He confiscated all the available food stores and set up kitchens to feed the survivors and organized parties to toss the rotting corpses of the dead into the ocean with minimal religious services, so their bodies would not spread disease.

Over the next 22 years, Carvalho was the de facto dictator of Portugal, doing everything he could to modernize the country and rebuild the city in a modern and seismically safe fashion. He also used his power to destroy the influence of the Jesuits on society and to banish several noble houses when they plotted against him and the king. He was eventually promoted to the rank of Marquês de Pombal (the name for which he is best known to history), and more than any man, brought Portugal out of the Middle Ages and into the Enlightenment. When the king died in 1777, Carvalho lost his authority. He was banished by the queen, who favored the Jesuits and restored many of the rich families that Carvalho had persecuted. Disgraced by the royal family, Carvalho died of old age in his country estate in 1781, but soon thereafter, Portugal was swallowed up by Napoleon's empire, and the royal family and their reactionary ways were swept aside by the new political movement. Today, Carvalho (or Pombal) is viewed as one of Portugal's greatest heroes, and his reputation has been completely restored.

The earthquake was felt not only in Lisbon (where at least 60,000 and maybe as many as 100,000 people died) but all over the region, especially North Africa, Spain, and France. More than 10,000 died in Morocco as well. It shocked European sensibilities, because there had not been an earthquake this large in historic memory. There were great earthquakes in antiquity, of course, and there were a number of smaller earthquakes in Italy and Greece over the centuries. Huge earthquakes had been felt by Europeans visiting exotic places such as Peru or Haiti over the past century, but no severe earthquake had struck western Europe in recorded history. The Lisbon earthquake seemed unprecedented and much like an act of God.

For this reason, it had a profound effect on the educated people of Europe as well. Before 1755, one of the dominant schools of thought in Enlightenment philosophy was called *optimism*. Epitomized by the writings of Gottfried Wilhelm von Leibniz, this philosophy minimized the problem of evil and pain. If the universe is the creation of an all-wise, all-powerful, perfect Creator, then pain, suffering, and evil are just part of His overall plan. We cannot scrutinize or understand the ways of an all-powerful, all-knowing God, so we should not question His plan or try to second guess Him about why he allows evil to exist. We should not worry about why and how evil occurs but celebrate the infinite goodness and wisdom of the Creator. In the famous lines of Alexander Pope's *An Essay on Man* (1733):

All Nature is but Art, unknown to Thee;
All Chance, Direction, which thou canst see;
All Discord, Harmony, not understood;
All partial Evil, Universal Good;
And, in spite of Pride, Erring Reason's spite,
One truth is clear: WHATEVER IS, IS RIGHT.

The Lisbon earthquake deeply shocked most of Europe and especially the optimist school of thought. Unlike more worldly and sinful cities, Lisbon was one of the most devout, so it was hard to believe that it merited special punishment from God. Many Protestants thought that the evils of the Inquisition were the cause for divine punishment, but Lisbon was not unique in this regard. Madrid and many other cities in the Catholic nations also had equally virulent forms of the Inquisition.

Yet other philosophers and thinkers, such as Immanuel Kant, Jean-Jacques Rousseau, and especially Voltaire, saw the Lisbon earthquake as a direct repudiation of the entire optimistic worldview. In his classic work *Candide*, Voltaire uses the character Dr. Pangloss to mouth the ideas of Leibniz: this is the best of all possible worlds, so everything is for the better, even if it appears evil at the time. Voltaire satirized this notion by having Dr. Pangloss spout such absurdities as "noses were made for spectacles" and "legs were made for breeches." The main character, Candide, is an innocent youth who suffers through an enormous range of calamities, yet never questions Dr. Pangloss's teaching that this must be the best of all possible worlds and part of God's plan. Naturally, Candide and Dr. Pangloss arrive in Lisbon just as the great earthquake strikes:

They had scarcely set foot in the town [Lisbon] when they felt the earth tremble under their feet; the sea rose in foaming masses in the port and smashed the ships which rode at anchor. Whirlwinds of flame and ashes covered the streets and squares; the houses collapsed, the roofs were thrown upon the foundations, and the foundations were scattered; thirty thousand inhabitants of every age and sex were crushed under the ruins . . .

20   Catastrophes!

"What can be the sufficient reason for this phenomenon?" said Pangloss.
"It is the Last Day!" cried Candide. (Voltaire, *Candide*, or *Optimism*)

As if experiencing this great earthquake were not enough, Candide and Pangloss are soon accused by the Inquisition of causing the quake and sentenced to death by burning at the stake.

After the earthquake which destroyed three-quarters of Lisbon, the wise men of that country could discover no more efficacious way of preventing total ruin than by giving the people a splendid *auto-da-fe* ["act of faith," the term the Inquisition applied to their executions of heretics]. It was decided by the university of Coimbra that the sight of several persons being slowly burned in great ceremony is an infallible secret for preventing earthquakes . . .

They marched in procession and listened to the most pathetic sermon, followed by lovely plainsong music. Candide was flogged in time to the music, while the singing went on; the Biscayan and the two men who had not wanted to eat bacon [implying they were Jewish] were burned, and Pangloss was hanged, although this is not the usual practice. The very same day, the earth shook again with a terrible clamor.

Candide, terrified, dumbfounded, bewildered, covered with blood, quivering from head to foot, said to himself: "If this is the best of all possible worlds, what are the others?" (Voltaire, *Candide*, or *Optimism*)

Candide escapes only to go on to his next misfortune, while Pangloss does not. For reasons not explained by Voltaire, Pangloss somehow reappears later in the story. This masterpiece of Western literature, through its vicious satire, exposed the "optimism" of the day for the naïve, empty, and witless notion that it was.

For the faithful, of course, earthquakes remained evidence of God's punishment, but the Lisbon earthquake also bore the first fruits of modern seismology. Carvalho sent out a questionnaire in 1760, which attempted to get an accurate survey from the survivors about the motions of the ground and sea, as well as the reaction of the people, at wide distances from the earthquake. It was one of the first naturalistic, rational, nonsupernatural attempts to understand earthquakes. In 1760, the Reverend John Michell, a professor of geology and mineralogy at Cambridge in England, published the results of his research as *Conjectures concerning the Cause of Observations upon the Phaenomena of Earthquakes*, which can be considered the first publication in modern seismology. He not only compiled all sorts of observations about the Lisbon earthquake and many others but also correctly realized that earthquakes were caused by energy waves propagating along the earth's surface, which he analogized to snapping the edge of a large carpet on the floor.

In recent years, the likely cause of the 1755 Lisbon earthquake has been determined. There are a number of large faults that run east-west through the Iberian Peninsula and North Africa and out onto the Atlantic seafloor. These faults are part of the boundary between the Eurasian and African plates. Indeed, the entire Mediterranean is fractured by faults that are part of this system, causing the earthquakes that occur in Italy and Greece over the centuries. The African plate continues to push northward into southern Europe, causing the uplift of the Alps. Based on the time of the arrival of the tsunamis (which travel at speeds of 400–500 km/h), the epicenter of the 1755 quake had to be on the faults about 200 to 250 km southwest of Lisbon, near the Gorringe Bank or Ampere Bank on the eastern Atlantic seafloor. These faults are still active. Hundreds of earthquakes emanated from them between 1930 and 1980; five were magnitude 7.0 or above. The 1969 earthquake in this region caused extensive damage in Lisbon again (Mercalli intensity of VII). On the basis of the described ground effects of the 1755 quake, it is estimated to have had a Mercalli intensity of IX and a Richter magnitude of 8.5.

The Birth of Modern Seismology: San Francisco, 1906

In 1906, San Francisco was a bustling, bawdy place, the biggest city and the commercial hub of the West Coast. Since the 1849 Gold Rush, its population had exploded to more than 400,000 people and was enriched not only by the mining in the Sierras but also by the growth of agriculture and fishing all around San Francisco Bay. San Francisco was also notoriously corrupt; many city officials were known to take bribes and kickbacks. The infamous Barbary Coast along the waterfront was crowded with saloons and brothels, and in Chinatown, there were opium dens with prostitution and slavery.

San Francisco was no stranger to destructive earthquakes and fires. During the Gold Rush days of 1849–1851, it had almost burned down six times, because most of the buildings were flimsy wood and canvas structures, hastily thrown together to serve the booming population. After the sixth big fire in 1851, construction shifted to brick and stone buildings, and the fire department became better organized and equipped. Huge cisterns of water were placed strategically around the city for the fire department to draw from. Consequently, the city had not suffered a catastrophic fire in fifty-five years, but because of rampant corruption, the fire department was underfunded and run by cronyism; therefore, the cisterns had been neglected. There had also been large tremors in 1836, in 1838 (when it was still a sleepy Mexican pueblo), and in 1868, twenty years after the Gold Rush started. In his travelogue of his adventures in the West entitled *Roughing It* (1871, 314), Mark Twain wrote of his experience in San Francisco in 1868:

> There came a really terrific shock . . . and there was a heavy grinding noise as of brick houses rubbing together . . . As I reeled about on the pavement trying to keep my footing, I saw a sight! The entire front of a tall four-story brick building . . . sprung outward like a door and fell sprawling across the street . . . Every door of every house . . . was vomiting a stream of human beings; and . . . there was a massed multitude of people stretching in endless procession down every street.

Thirty-eight years had erased the memories of this event from the minds of most San Franciscans by 1906. There had been so many new arrivals to the city that almost none had lived in the city in 1868 and most were not even born then. The toppled buildings had long ago been rebuilt, and many more buildings had gone up since then.

The night of Tuesday, April 17, was unseasonably warm. Many affluent citizens had attended a performance of *Carmen* at the Grand Opera House, starring the legendary tenor Enrico Caruso in the role of Don José. (Ironically, the newspapers of the day featured accounts of the eruption of Mount Vesuvius in Italy, not far from Caruso's hometown of Naples.) Dressed in their finest clothes, the opera patrons had headed home and gone to bed.

The streets were quiet when, at 5:12 a.m., the entire city awoke to a series of sharp jolts. One policeman on duty that morning described the ground movement as a wave in a rough sea, rolling down the street. Officer Jesse B. Cook was standing at the eastern end of Washington Street and was one of the first to witness the waves of energy and water approaching from the north. Not only were there waves of water advancing down the street, but the entire street was undulating. The buildings and pavement were lifted up and then toppled over. His report includes the following description:

> The earth seemed to rise under me, and at the same time both Davis and Washington streets opened up in several places and water came up out of these cracks. The street seemed to settle under me, and did settle in some places from about one to three feet. The buildings around and about me began to tumble and fall and kept me pretty busy for a while dodging bricks. I saw the top story of the building at the southwest corner of Washington and Davis fall and kill Frank Bodwell.

A night clerk at the Valencia Street Hotel (fig. 1.2D) ran from the building and gave this account: "The hotel lurched forward as if the foundation were dragged backward from under it, and crumpled down over Valencia Street. It did not fall to pieces and spray itself all over the place, but telescoped down on itself like a concertina" (*Morning Edition* 2006).

The people who were on the first three floors were crushed to death (killing at least 100), while those who happened to have rooms on the fourth floor simply stepped out onto the street. Another witness named P. Barrett wrote: "We could not get to our feet. Big buildings were crumbling as one might

Fig. 1.2. A, The fires sweeping through ruins of San Francisco, viewed by the survivors from the Golden Gate Heights. B, The ruins of City Hall. C. Streets, sidewalks, and rail lines all distorted by the ground motion, along Ninth Street between Bryant and Brannan streets. D, The four-story Valencia Street Hotel pancaked down to ground level and killed many, then burned when the fire swept through after this photo was taken. E, Most buildings, like the California Hotel, were reduced to rubble. (Photos courtesy USGS Photo Library)

crush a biscuit in one's hand. Ahead of me a great cornice crushed a man as if he were a maggot" (*Morning Edition* 2006).

The seven-story Palace Hotel was the grandest hotel in San Francisco, housing celebrities such as Caruso, presidents, and kings. With more than 800 rooms and four newfangled elevators, the Palace was the largest hotel in the country. It also had 700,000-gallon iron water tanks built under the roof in case of fire. When the shaking started, the horses in the carriage entrance bolted and the trees swayed, but otherwise the building held up. Caruso had only gone to bed two hours earlier after his post-performance meal, but he was severely shaken and panic-stricken. Different versions of how he reacted to the events have been published, but one account says that he immediately left town after he had put on a fur coat over his nightclothes, muttering "'ell of a town! 'ell of a town! I never come back!" And he never did.

Harry Walsh, another policeman, had witnessed the death and destruction and saw huge cracks open in the pavement on Fremont Street, which closed and reopened as the shock waves passed. Then he saw a herd of longhorn cattle stampeding toward him along Mission Street from the direction of the docks. Apparently, they had just been unloaded from an inbound ship and were being driven to the stockyards south of town when the shock occurred. The Mexican *vaqueros* had fled in panic, leaving their cattle to run through the streets of the city. As Walsh wrote:

> While a lot of them were running along the sidewalks of Mission Street, between Fremont and First streets, a big warehouse toppled onto the thoroughfare and crushed most of them clean through the pavement into the basement, killing them and burying them outright. The first that I saw of the bunch were caught and crippled by falling cornices, or the like . . . and were in great misery. So I took out my gun and shot them. Then I had only six shots left, and I saw that more cattle were coming along, and that there was going to be big trouble.
>
> At that moment, I ran into John Moller, who owned the saloon . . . I asked him if he had any ammunition in his place, and if so, to let me have some quick. He was very scared and excited over the earthquake and everything; and

when he saw the cattle coming along, charging and bellowing, he seemed to lose more nerve.

Anyway, there was no time to think. Two of the steers were charging right at us while I was asking him for help, and he started to run for his saloon. I had to be quick about my part of the job because, with only a revolver as a weapon, I had to wait until the animal was quite close before I dared fire. Otherwise, I would not have killed or even stopped him.

As I shot down one of them I saw the other charging after John Moller, who was then at the door of his saloon and apparently quite safe. But as I was looking at him and the street, Moller turned, and seemed to become paralyzed with fear. He held out both hands as if beseeching the beast to go back. But it charged on and ripped him before I could get near enough to fire. When I killed the animal it was too late to save the man . . .

Then a young fellow came running up carrying a rifle and a lot of cartridges. It was an old Springfield and he knew how to use it. He was a cool shot, and he understood cattle, too. He told me he came from Texas . . . we probably killed fifty or sixty. (*Morning Edition* 2006)

Although the shaking lasted approximately 40 seconds, the devastation was complete (fig. 1.2B–1.2E). Nearly every unreinforced masonry building in the city had fallen down, although most of the wood- and steel-frame buildings were still standing. At first, the city was quiet, with a huge cloud of dust obscuring much of the damage. Soon, an even more destructive force was brewing. With all the oil-burning lamps, candles, and fireplaces toppled, the many wooden structures were kindling for fires that grew out of control (fig. 1.2A). These fires raged for almost four days, burned more than 28,000 structures, and leveled more than three-quarters of the city, causing 10 times as much damage as the earthquake. Most of the city's water supply was transported in rigid iron pipes, and 30,000 pipes ruptured with the intense shaking. Therefore, there was almost no water for the firefighters to use. In addition, the city's fire chief was killed when his fire station collapsed, so the firefighters were leaderless. They tapped into long-neglected cisterns that had been set up for

ırpose, but because of the scale of the citywide blazes, that water was
ıusted.

In desperation, the firefighters and the 2,000 federal troops that had arrived (without even a declaration of martial law) began to use dynamite to create a firebreak by blowing up buildings in the path of the fire.

"One of the problems was the type of explosives that they used," according to Philip Fradkin, author of *The Great Firestorms of 1906.* "Gunpowder is flammable and spreads fire. And they made the mistake on the end of the second day of dynamiting a huge chemical warehouse . . . and that was just pyrotechnics plus" (*Morning Edition* 2006).

By this point, a huge area of the city was ablaze. The mayor issued a "shoot-to-kill" order for anyone found looting. Thousands of residents wore layers of their best clothes and walked to the ferryboats to cross San Francisco Bay to Oakland or to tent camps scattered throughout the city.

A single woman, Rosa Barreda, who lived with her mother, described the horror in a letter to a friend:

> Many burned-out people passed our house with bundles and ropes around their necks, dragging heavy trunks. From the moment they heard that fatal, heart-rending sound of the trumpet announcing their house would be burned or dynamited, they had to move on or be shot. As the sun set, the black cloud we watched all day became glaringly red, and indeed it was not the reflection of our far-famed Golden Gate sunset. (*Morning Edition* 2006)

Even though the earthquake had not damaged all the neighborhoods, those that were spared the seismic damage were later burned to the ground. The official death toll in San Francisco was about 700, although the Chinese and Mexican laborers were undercounted. However, the damage was felt widely up and down the region, from Northern California to Los Angeles to Nevada, and especially in the southern Bay Area cities such as San Jose. There were about 3,000 to 5,000 dead. Half of the population of 400,000 was left homeless.

As the catastrophe ended, the city boosters began a publicity campaign to de-emphasize the earthquake's role and referred to the event as the San Francisco Fire of 1906. They didn't want to discourage out-of-town visitors and investors who would consider a fire familiar and manageable but might be frightened off by the prospect of another earthquake. Instead, these boosters spread false stories in the press about how the city had risen from the ashes like a phoenix by the following weekend. Others felt that emphasizing the earthquake made the situation seem hopeless, whereas blaming it on a human-made fire disaster on a badly built city suggested that humans could also solve the problem.

Geologists found this act of denial both humorous and frustrating. In 1908, John Branner wrote the following in the *Bulletin of the Seismological Association of America* (an organization that was established because of the 1906 earthquake):

A major obstacle to the proper study of earthquakes [was] the attitude of many persons, organizations, and commercial interests to the false position that the earthquakes are detrimental to the good repute of the West Coast, and that they are likely to keep away business and capital, and therefore the less said about them the better. This theory has led to the deliberate suppression of news about earthquakes, and even the simple mention of them.

Shortly after the earthquake of April 1906 there was a general disposition that almost amounted to concerted action for the purpose of suppressing all mention of that catastrophe. When efforts were made by a few geologists to interest people and enterprises in the collection of information in regard to it, we were advised and even urged over and over again to gather no such information, and above all not to publish it. "Forget it," "the less said, the sooner mended," and "there hasn't been any earthquake" were the sentiments we heard on all sides.

There is no doubt about the charitable feelings and intentions of those who take this view of the matter, and there is reasonable excuse for it in the popular but erroneous idea prevalent in other parts of the country that earthquakes are all terrible affairs; but to people interested in science, it is not necessary to say that such as attitude is not only false, but it is most unfortunate, inexcusable, untenable, and can only lead, sooner or later, to confusion and disaster. (Branner 1913)

As soon as the fires were out, the city immediately began to rebuild. Only nine years after the great earthquake, San Francisco hosted the Panama-Pacific International Exhibition, a spectacular world's fair. It was built by pouring the earthquake debris into the bay to create a new landfill, now known as the Marina District. Ironically, this soft bay fill was even more susceptible to shaking than the bedrock of the rest of the city, and much of the damage in the 1989 Loma Prieta earthquake occurred in the Marina District, where houses sank into the soft ground before collapsing and burning.

In addition to stimulating construction, the earthquake also fostered one of the first scientific commissions to study all aspects of an earthquake and its possible causes. The State Earthquake Investigation Commission (SEIC) was set up under the leadership of Andrew C. Lawson, a legendary geologist and a faculty member at the University of California, Berkeley. (One of the distinctive minerals of the California Coast Ranges is named *lawsonite* in his honor.) Lawson and his commission studied nearly every aspect of the great earthquake and produced a pair of volumes, coedited with the pioneering seismologist Harry Fielding Reid of the Johns Hopkins University, entitled *The Report of the State Earthquake Commission*, published in 1908 with more than 300 pages.

The commission and the coauthors of the volume included many of the leading lights of geology and geophysics at that time, including Grove Karl Gilbert of the U.S. Geological Survey, whose legendary discoveries are still impressive a century later. Gilbert happened to be in Berkeley when the quake struck, and although he couldn't get into San Francisco for days, he traveled up and down the region, mapping and photographing the fault displacement from near Tomales Bay down to the southern Bay Area and San Juan Bautista. H. F. Reid used the 1906 earthquake to develop the elastic rebound theory of seismology, which is still accepted today. J. C. Branner (quoted earlier) was the founder of the Stanford Geology Department and later president of Stanford University. H. O. Wood later founded the Seismological Lab at Caltech. F. E. Matthes mapped the topography of many places, including Yosemite and other national parks. G. Davidson was the first president of the Seismological Society of America. Fusakichi Omori was one of the most-celebrated seismologists in Japan.

This all-star lineup of geologists and seismologists compiled an impressive report, not only documenting the effects of the earthquake on structures and how it felt but also detailing the physical effects, especially the fault offset and ground rupture (fig. 1.3). Before the report, geologists did not all accept that faults caused earthquakes, but the 1908 report settled that question once and for all. The commission traced the rupture of the fault all the way from Point Delgada north of Tomales Bay, down to San Juan Bautista, and even mentioned effects as far south as Whitewater Canyon near San Bernardino. Although they didn't connect all these activities to the same San Andreas fault, or recognize that most of its motion was horizontal strike-slip rather than vertical offset, they recognized the California Coast Ranges were permeated by faults, including the Hayward fault beneath Oakland and the San Jacinto fault west of Palm Springs. They also documented many related phenomena, such as coseismic landslides and uplifted and downdropped crustal rocks, and made maps of damage to structures, descriptions and analyses of the few available seismographs, and accounts of previous California earthquakes. Part 2 of the report, authored mainly by Reid, not only proposed the elastic rebound theory of earthquakes but also laid the foundation for the geophysics of earthquakes that is still used. In short, the SEIC report was the foundation for all of modern seismology.

The most widely accepted estimate for the magnitude of the earthquake is a moment magnitude (M_w) of 7.8; however, other values have been proposed, from 7.7 to as high as 8.3. The main shock epicenter occurred offshore about 2 miles (3 km) from the city, near Mussel Rock. It ruptured along the San Andreas fault both northward and southward for a total length of 296 miles (477 km). In 2006, the geological community commemorated the centennial

Fig. 1.3. Ground effects of the 1906 Bay Area earthquake documented by G. K. Gilbert. A, Fault scarp near Olema, California, with Mrs. Gilbert for scale. The block to the left has shifted away from the viewer into the distance relative to the block on the right. B, A fence near Woodville, California, offset by the fault motion. The segment in the foreground shifted 3 m (9 feet) to the left relative to the background during the earthquake. (Photos by G. K. Gilbert, courtesy USGS Photo Library)

of the San Francisco earthquake with numerous scientific meetings and publications. The overwhelming message from the assembled scientists was that the danger was not past. The 1989 Loma Prieta earthquake occurred on a stretch of the San Andreas fault just south of the 1906 break and caused severe damage in San Francisco again (including breaking spans of the Bay Bridge and causing many houses in the Marina District to sink into the ground and burn). The major faults along the East Bay, especially the Hayward fault, are considered even more dangerous. There's even an outstanding video on the Internet documenting the 1906 event and discussing the next great earthquake in the region, along with many other important resources regarding earthquakes at (www.1906eqconf.org/). Even though more than a century has passed since the events of 1906, Bay Area residents should not assume that there are no more earthquakes in their future.

Seismology Confirms Plate Tectonics: Alaska, 1964

If the 1755 Lisbon quake first brought earthquakes out of the realm of superstition, and the 1906 San Francisco quake led to modern seismology, the earthquake that hit Alaska on Friday, March 27, 1964, placed seismology within the realm of our modern plate tectonic understanding of the earth. Ironically, this day happened to be Good Friday, when, according to the Gospel of Matthew 27:51, "and the earth shook and the rocks were split" at Jesus's crucifixion.

The shaking started at 5:36 p.m. in Alaska, just as people were headed home for the weekend, and many were in church for Good Friday services. Luckily, that meant few people were still at work in buildings in downtown Anchorage, and many businesses were already closed. Tides were low and fishing season was over, so there were relatively few people on the docks. The weather was relatively warm and headed into spring. Many people were outdoors and did not have to hide indoors for the winter. In addition, Alaska was sparsely populated in 1964, with only 253,000 people. If the quake had struck a heavily populated area or at a time when more people were indoors, more people would have been killed than the 131 who died.

The earthquake started as a low rumbling, and then the entire region began to shake violently and kept shaking for a full five minutes. Witnesses reported that it seemed like an eternity. By contrast, the shaking in the Loma Prieta earthquake of 1989 or the 1994 Northridge quake lasted only 30 seconds. In fact, the Alaska earthquake was not a single sharp shift of the crust, as in most earthquakes, but a series of back-to-back slips along the fault at 19, 28, 29, 44, and 72 seconds after the initial shock, which meant that the shock waves kept propagating over and over, and it took 5 minutes for all that energy to die down. The most striking effect was the damage to buildings in such cities as Anchorage, where entire streets were ripped open and dropped down 3 m (11 feet) in response to the land slipping beneath them (fig. 1.4A, 1.4B). Down at the port towns of Valdez and Seward (fig.1.4C), a huge tsunami 10 m (35 feet) high swept over the docks and drowned 122 people in a matter of minutes. (The tsunami was responsible for the most deaths, 122 of the 131 lives lost for the entire earthquake.) An oil tanker docked in Seward wrenched loose from a pipeline, erupted in flames, and the fire spread to the nearby oil tanks, which promptly exploded. Burning oil on the water washed inland. In both ports, ships were battered against piers. The town of Valdez was so badly damaged that it was relocated to a new site.

Perhaps the most dramatic event of the earthquake was the catastrophic landsliding in the Turnagain Heights subdivision, just southeast of Anchorage. The bluff on which the development had occurred broke apart when the violent shaking caused the deeply buried layers of mud to liquefy. The ground then broke into a series of slump blocks that began sliding steadily toward the sea, carrying houses, cars, trees, and streets with them (plates 1A, 1B). Back when I was a kid in 1964, I vividly remember reading the eyewitness accounts in *National Geographic* magazine. One story was described by Tay Thomas, wife of TV producer and Alaska senator Lowell Thomas, Jr. (and includes their children, David, 6, and Anne, 8):

Seeing that fissure widen next to me was the exact picture I'd always had in my mind of what happened in a violent earthquake . . . Then our whole lawn broke up into chunks of dirt, rock, snow, and ice. We were left on a wildly bucking

Fig. 1.4. A, Collapse of Fourth Avenue near C Street in Anchorage due to a landslide caused by the earthquake. Before the shock, the sidewalk on the left, which is in the graben, was at street level on the right. The graben subsided 3 m (11 feet) in response to 3.5 m (14 feet) of horizontal movement. B, Wreckage of the J.C. Penney Department Store at Fifth Avenue and D Street in Anchorage. The building failed after sustained seismic shaking. Most of the rubble had been cleared from the streets. C, Earthquake subsidence, fire, and tsunami damage to the Seward port facility. (Photos courtesy USGS Photo Library)

slab; suddenly it tilted sharply, and we had to hang on to keep from slipping into a yawning chasm . . . I had the weird feeling that we were riding backward on a Ferris wheel, going down. I always hated riding on them anyway . . .

We started walking to the right, staying far enough away from the cliff to avoid still-falling sand. It was then that I first noticed Dr. Perry Mead's house — he was our next-door neighbor. Nothing showed but the flat roof. I could see two of their little children standing on top of a car . . . they would probably be safer standing on that car roof, I thought, than scrambling among the rocks and crevasses.

A man appeared above the cliff . . . and he shouted down that he would find some rope. (Thomas 1964, 144–45)

Once the shaking had stopped, the entire neighborhood was torn apart and uninhabitable, with millions of dollars in damage. The stories of huge fissures in the ground led to common myths (propagated in movies like *Superman* with Christopher Reeve) that earthquakes produce large chasms as the ground tears apart and lava comes up from below. In actuality, the chasms produced by the Alaska earthquake were the product of landslides and slumping triggered by the earthquake and not on the site of the actual fault line.

After the quake ended, geologists joined the relief efforts. They determined that the quake had a moment magnitude of 9.2, or a Richter magnitude of 8.4 to 8.6, making it the second-largest earthquake ever recorded, second only to the great Chile earthquake of 1960. The movement had occurred on a fault zone more than 400 km (250 miles) wide, and more than 110,000 square miles of seafloor were shifted up or down. The focal depth was 20–50 km below the epicenter along the Aleutian trench.

More important, U.S. Geological Survey geologists such as George Plafker documented amazing ground displacements as a result of the quake. In some regions, rocks that had been below low tide were suddenly uplifted 12 m (38 feet), stranding docks, beaches, and their sea life above the highest tides (plates 2A, 2B). In other cases, areas that had been above sea level sank down about 2.3 m (7.5 feet) and were permanently submerged. By mapping this subsidence and uplift, Plafker realized that the main motion of the fault had been a giant thrust, wherein one plate pushed under another in a process known as *subduction*. The rocks at the leading edge of the plate overlying the Aleutian trench had buckled upward, while those behind the upwarp (along the Kenai Peninsula and Cook Inlet, including Anchorage), had warped downward and sunk. Combined with the seismograph records, which showed that the entire fault motion was a giant thrust, geologists realized that the Alaska earthquake was the first good example of an active subduction zone, part of the newly proposed theory of plate tectonics. As the years passed, more pieces of the Alaska puzzle were fit into the plate tectonic paradigm. But Good Friday 1964 was the first time geologists could see subduction tectonics in action.

Intraplate Earthquakes: New Madrid, Missouri, 1811–1812

The 1964 Alaska earthquake was on a subduction zone plate boundary, while the California earthquakes were along faults that slide horizontally, known as a transform plate boundary. The 1755 Lisbon quake was part of the crush zone as Africa collides with Europe. Although 99 percent of all earthquakes occur near plate boundaries, some of the biggest ever recorded do not. Earthquakes are common in the Great Basin region, from the Wasatch fault that runs along Interstate 15 in central Utah, across the fault-block mountains of western Utah and Nevada, to the faults along the eastern face of the Sierra Nevada Mountains in California. New England has experienced a number of earthquakes over the years, especially along the St. Lawrence Valley. New York has been shaken by small earthquakes along the Ramapo fault. Charleston, South Carolina, was rocked by a huge earthquake in 1886 that toppled many of the historic antebellum brick buildings. Probably the largest earthquakes ever experienced in North America didn't occur on the East or West coasts but in the "stable" middle of the continent, near New Madrid, Missouri. Four major quakes, each followed by more than 1,800 aftershocks, occurred on December 16, 1811 (two separate quakes), January 23, 1812, and February 7, 1812. Because there were few towns of any size in the area at the time, the damage to buildings was slight, but the entire Missisippi River region from Memphis, Tennessee, to Cairo, Illinois, was rocked and churned by the seismic waves, turning the ground into quicksand and causing the Mississippi River to change course, where it eventually drowned the original site of New Madrid (fig. 1.5). However, the quakes were so strong that they were felt as far as Boston, where they rang church bells and toppled steeples. An eyewitness account written by Eliza Bryan in 1816 describes it vividly:

> On the 16th of December, 1811, about two o'clock, A.M., we were visited by a violent shock of an earthquake, accompanied by a very awful noise resembling loud but distant thunder, but more hoarse and vibrating, which was followed in a few minutes by the complete saturation of the atmosphere, with sulphurious vapor, causing total darkness. The screams of the affrighted inhabitants running to and fro, not knowing where to go, or what to do—the cries of the fowls and beasts of every species—the cracking of trees falling, and the roaring of the Mississippi—the current of which was retrograde for a few minutes, owing as is supposed, to an irruption in its bed—formed a scene truly horrible.
>
> From that time until about sunrise, a number of lighter shocks occurred; at which time one still more violent than the first took place, with the same accompaniments as the first, and the terror which had been excited in everyone, and indeed in all animal nature, was now, if possible doubled. The inhabitants fled in every direction to the country, supposing (if it can be admitted that their minds can be exercised at all) that there was less danger at a distance from, than near to the river. In one person, a female, the alarm was so great that she fainted, and could not be recovered . . .
>
> At first the Mississippi seemed to recede from its banks, and its waters gathering up like a mountain, leaving for the moment many boats, which were here on their way to New Orleans, on bare sand, in which time the poor sailors made their escape from them. It then rising fifteen to twenty feet perpendicularly, and expanding, as it were, at the same moment, the banks were overflowed with the retrograde current, rapid as a torrent—the boats which before had been left on the sand were now torn from their moorings, and suddenly driven up a little creek, at the mouth of which they laid, to the distance in some instances,

Fig. 1.5. Boatmen struggle to shore as the New Madrid earthquake transforms the Mississippi River into a rolling swell. (Painted by J. F. Jungling, in the State Historical Society of Missouri. Courtesy Pacific Earthquake Engineering Research Center Earthquake Engineering Online Archive.)

of nearly a quarter of a mile. The river falling immediately, as rapid as it had risen, receded in its banks again with such violence, that it took with it whole groves of young cotton-wood trees, which ledged its borders. They were broken off [with] such regularity, in some instances, that persons who had not witnessed the fact, would be with difficulty persuaded, that it has not been the work of art. A great many fish were left on the banks, being unable to keep pace with the water. The river was literally covered with the wrecks of boats, and 'tis said that one was wrecked in which there was a lady and six children, all of whom were lost.

In all the hard shocks mentioned, the earth was horribly torn to pieces—the surface of hundreds of acres, was, from time to time, covered over, in various depths, by the sand which issued from the fissures, which were made in great numbers all over this country, some of which closed up immediately after they had vomited forth their sand and water, which it must be remarked, was the matter generally thrown up. . . . We were constrained, by the fear of our houses falling, to live twelve or eighteen months, after the first shocks, in little light camps made of boards; but we gradually became callous, and returned to our houses again. Most of those who fled from the country in the time of the hard shocks have since returned home. We have, since the commencement in 1811, and still continue to feel, slight shocks occasionally. It is seldom indeed that we are more than a week without feeling one, and sometimes three or [four] in a day. There were two this winter past much harder than we had felt them for two years before; but since then they appear to be lighter than they have ever been, and we begin to hope that ere long they will entirely cease. (*Lorenzo Dow's Journal*, 1849, 344–46)

Why did such huge quakes occur so far from plate boundaries? For a long time, it was a mystery, but then seismologists began to record the constant activity of small quakes in the region and realized there was an ancient rift valley buried deep beneath the thick blanket of Mississippi River sediments. It first formed about 550 million years ago when central North America tore apart from other continents that adjoined it. The rift was then reactivated about 200 million years ago when eastern North America pulled away from Africa and South America as the supercontinent of Pangea broke up. Even though the valley is no longer rifting apart, the buried faults are still active and apparently responding to crustal stresses that compress across all of North America. The New Madrid fault system is, in many ways, even more dangerous than well-exposed faults like the San Andreas in California. Unlike exposed faults with a good surface record that can be carefully studied, the deeply buried New Madrid fault system cannot be studied at the surface, and our only information comes from the many small earthquakes at great depth that seismographs detect.

What Causes Earthquakes?

Cultures around the world have had different explanations for earthquakes over the millennia. Usually they were blamed on the anger of the gods, especially in seismically active regions around the Mediterranean and the Middle East. In Japanese folklore, a giant catfish named Namazu lived in the mud below the surface of the ground and would shake the earth if the gods did not restrain him. In India, the earth shook when one of the eight elephants that bore it on their backs decided to move. In Siberia, the earth shook when the dog hauling the earth in its sled decided to scratch. Indigenous American tribes of the West Coast thought that arguments between the turtles that held up the earth caused it to shake. As we already saw at the beginning of the chapter, the anger of God was used to "explain" the Lisbon earthquake, but the Lisbon earthquake coincided with the beginnings of the Scientific Revolution in 1755, which led to the first nonmythic explanations of earthquakes.

Earthquakes are a fact of life on the earth's surface. More than 1 million detectable earthquakes happen each year, and in seismically active regions such as Southern California, there are small earthquakes every few minutes, even though most people do not feel them. However, gigantic earthquakes that kill many people only occur a few times in a decade around the world, and these are the ones that we hear about in the news. Of these quakes, nearly all are concentrated on plate boundaries (fig. 1.6). The most active of these is the "Ring

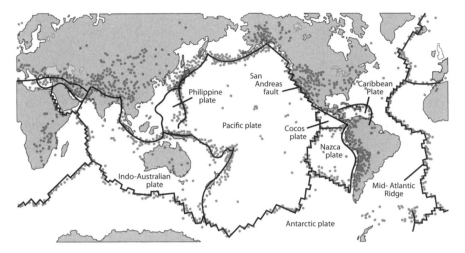

Fig. 1.6. Map of world seismicity (dots), showing that most earthquakes occur along plate boundaries. (Courtesy USGS, redrawn by Pat Linse)

of Fire" around the Pacific Rim, where many subduction zones and transform faults cause constant seismic activity on the Pacific Coast of the Americas, the Aleutians, Kamchatka, Japan, the Philippines, Malaysia and Indonesia, New Guinea, and New Zealand. The second major belt of seismic activity runs along the region where Africa and India have been plowing northward into the belly of Eurasia, causing the great seismic zones that run from China and Southeast Asia through the Himalayas to Afghanistan, Iran, the Caucasus, Turkey, and then westward across to the Alps and northern Mediterranean. These two seismic regions account for nearly all of the world's great earthquakes and most of the deaths as well. In these places, two great crustal plates are pushing together, with one of them subducting beneath the other (the subduction zones of the world's great oceanic trenches), or colliding and crumpling like two huge cars in a fender bender (as in the Himalayas or the Alps). The great midocean ridges, where the crust of the earth pulls apart, are subject to many small earthquakes. These result from the release of strain as oceanic crust tears open and allows magma to well up from the mantle. They are felt only by the most sensitive seismographs. A small number of earthquakes occur in midcontinental regions away from the plate edge, such as the New Madrid or Charleston earthquakes.

The 1755 Lisbon earthquake was the first event in which the scientific notion of seismic waves was proposed. The 1906 San Francisco event conclusively showed that faults caused earthquakes. It also showed that the deforming crust on each side of a sliding fault snaps back after releasing tension. This is a mechanism known as *elastic rebound*. After the rocks along the fault snap back, they release a huge amount of energy, which propagates as waves away from the source of the earthquake, known as the *focus*, or *hypocenter*. (The spot on the map immediately above this deep focus is known as the *epicenter*.) The waves travel outward in big spherical pulses away from the source, as sound waves travel from a sound source or ripples from a disturbance in a pond. Each earthquake generates a number of different kinds of waves, named after their arrival on the seismogram. The first to arrive is the P, or primary wave, which travels about 5.5 km/sec (12,300 miles per hour [mph]) in crustal rocks, but only 1.5 km/sec (3,355 mph) in water. Its motion resembles a sound wave, with pulses of compressed and stretched out air molecules moving in the direction in which the wave travels. Next to arrive is the S, or secondary wave, which travels about 3.0 km/sec (6,711 mph) in rocks but does not travel through water. The energy of the S wave moves in an up-and-down shearing motion, similar to picking up the edge of a carpet and giving it a strong up and down shake, with ripples propagating across the carpet.

P waves and S waves travel not only along the ground but also through the entire body of the earth, so they are known as body waves. The same earthquake will also produce slower, lower-energy waves that travel only along the earth's surface. One wave, known as a Love wave, shifts the ground in a side-to-side motion like that of a sidewinder snake. The other, known as a Rayleigh wave, causes the ground to shift in small circular rotations up and down, like the water particles in a water wave. These waves can travel up to 5 km/sec and as slowly as 1 km/sec. Even though they are slower, they often have a much larger effect on the ground motion when they arrive, shaking buildings that have already been damaged by the earlier arrivals of the P and S waves. In fact, unless you are close to the focus, the P wave pulse is often missed by humans who are not sitting still and quiet during a quake, and the S waves cause the first shaking that gets people's attention.

How Do We Measure Earthquakes?

The fuller understanding of earthquakes could not occur until there was a reliable method of recording their occurrences, a device known as a seismograph. The first modern seismographs were invented in 1889. They are based on the idea that a heavy pendulum or weight suspended on a spring would stay in one position by inertia while the frame of the seismograph (and the entire earth) vibrates around it. This motion is transferred to a rotating drum of paper with a long pen, giving the familiar wiggly-line trace of seismographs. Even though it appears to us that the pen is wiggling up and down as the drum rotates, in reality the pen (and the heavy weight attached to it) is holding still, and the seismograph (the ground and you) are moving up and down. The seismogram, or the paper record of the seismograph, gives the time of the arrival of each wave generated by the same earthquake, and the highest "kick" of the pen gives the maximum amplitude of the waves felt by the seismograph. In recent years, the old paper drum seismographs have been replaced by a large magnet that moves up and down on a spring within a coil that senses the motion electronically, so the signal is now entirely an electronic pulse stored in a computer. Except for demonstration purposes and museum displays, seismologists no longer use the old drum and paper method.

For many years before the invention of modern seismographs, the only way to estimate the size of an event was to describe the ground motion, damage, and other ways in which it is felt. This led to the old Mercalli scale of intensity, proposed in 1902 by the Italian scientist Giuseppe Mercalli (table 1.1). The Mercalli intensity, ranked with Roman numerals from I to XII, goes from a Mercalli I ("detected only by seismic instruments; no damage") to a XII ("earthquake waves cause visible undulations of the ground surface; objects are thrown up off the ground; there is complete destruction of buildings and bridges of all types"). Such a scale is useful for earthquakes that happened

Table 1.1. Modified Mercalli Intensity Scale

Rank	Description
I	People do not feel any Earth movement.
II	A few people might notice movement if they are at rest or on the upper floors of tall buildings.
III	Many people indoors feel movement. Hanging objects swing back and forth. People outdoors might not realize that an earthquake is occurring.
IV	Most people indoors feel movement. Hanging objects swing. Dishes, windows, and doors rattle. The earthquake feels like a heavy truck hitting the walls. A few people outdoors may feel movement. Parked cars rock.
V	Almost everyone feels movement. Sleeping people are awakened. Doors swing open or close. Dishes are broken. Pictures on the wall move. Small objects move or are turned over. Trees might shake. Liquids might spill out of open containers.
VI	Everyone feels movement. People have trouble walking. Objects fall from shelves. Pictures fall off walls. Furniture moves. Plaster in walls might crack. Trees and bushes shake. Damage is slight in poorly built buildings. No structural damage.
VII	People have difficulty standing. Drivers feel their cars shaking. Some furniture breaks. Loose bricks fall from buildings. Damage is slight to moderate in well-built buildings; considerable in poorly built buildings.
VIII	Drivers have trouble steering. Houses that are not bolted down might shift on their foundations. Tall structures such as towers and chimneys might twist and fall. Well-built buildings suffer slight damage. Poorly built structures suffer severe damage. Tree branches break. Hillsides might crack if the ground is wet. Water levels in wells might change.
IX	Well-built buildings suffer considerable damage. Houses that are not bolted down move off their foundations. Some underground pipes are broken. The ground cracks. Reservoirs suffer serious damage.
X	Most buildings and their foundations are destroyed. Some bridges are destroyed. Dams are seriously damaged. Large landslides occur. Water is thrown on the banks of canals, rivers, and lakes. The ground cracks in large areas. Railroad tracks are bent slightly.
XI	Most buildings collapse. Some bridges are destroyed. Large cracks appear in the ground. Underground pipelines are destroyed. Railroad tracks are badly bent.
XII	Almost everything is destroyed. Objects are thrown into the air. The ground moves in waves or ripples. Large amounts of rock may move.

Source: From Federal Emergency Management Agency.

before the invention of seismographs, but it has severe limitations. The most obvious is that it highly depends on what kinds of buildings and structures are in the region, so in a rural area, there is relatively little damage even if the shaking is severe. As you get farther from the source of the shaking, the Mercalli intensity decreases, so it is hard to give a precise estimate of the true strength of the quake.

More significant, the damage on buildings and structures depends not only on the strength of shaking but especially on the type of ground on which the buildings rest. Loose sediment, especially if it is saturated with groundwater, may vibrate for much longer and more violently, as a bowl full of gelatin does when you shake it. In comparison, buildings sited on hard bedrock shake once when the first waves pass through and then stop vibrating. A vivid demonstration of this was the Mexico City earthquake of September 19, 1985. The actual site of the quake was on the Pacific Coast of Mexico, more than 350 km from Mexico City, where there were a small number of casualties. When the waves reached Mexico City, they shook the ancient lake-fill sediments that had once been lakebeds when the Aztecs ruled Mexico, and the loose liquefied sediment actually amplified the ground motion. This shook down more than 500 buildings and resulted in more than 9,000 deaths and $4 billion in damage (plate 3). A similar phenomenon occurred during the Northridge quake in California in 1994. I vividly remember feeling the shock waves pass quickly and then end in our house in Sunland, built on hard bedrock and yet only 10 km (17 miles) away. But the soft sediments of the San Fernando Valley vibrated like Jell-O and shook down many buildings, as did the soft liquefied sediment in Santa Monica, more than 15 km (24 miles) away. Yet towns such as Sherman Oaks, on the hard bedrock of the Santa Monica Mountains much closer to the epicenter, vibrated much less than the more distant areas of Santa Monica. Thus, the degree of ground shaking as an estimate of earthquake energy is highly biased by the nature of the structures and the ground on which they are built.

Modern seismographs allowed a much more reliable measurement of earthquake energy, starting with the 1935 scale proposed by Charles F. Richter of

Caltech. In this method, the highest peak of the seismogram is measured as a proxy for the maximum energy of the earthquake and then corrected for the distance from the earthquake to give a Richter magnitude. This scale is purely based on seismic intensity and corrects for features like distance from earthquake and differences in substrate. In recent years, seismologists have adopted a scale known as the moment magnitude (M_w), which corrects for the highly variable results the Richter scale gives on very large earthquakes. The moment magnitude uses not only the amplitude of the largest wave on each seismogram but also factors how much slip has occurred and the physical characteristics of rocks that broke during faulting. Under this system, some of the larger historical earthquakes give different Richter and moment magnitudes. The 1964 Alaska quake had a Richter magnitude of 8.4 but a moment magnitude of 9.2, while the 1906 San Francisco quake was demoted from a Richter magnitude of 8.3 to a moment magnitude of 7.9. The largest-known earthquake, the 1960 Chile quake, had a Richter magnitude of 8.5 but a moment magnitude of 9.5. Thus, when earthquakes are reported and compared, you need to be aware of the scale used to measure them. As we already showed, the Mercalli intensities can vary widely for earthquakes of the same Richter or moment magnitude, so a table that converts Mercalli intensity to Richter magnitude is misleading and incorrect.

Can We Predict Earthquakes?

Only fools and charlatans predict earthquakes.
— Charles F. Richter

As with any other natural disasters, people are eager for any kind of warning or prediction that would give them time to prepare. There have been many efforts to predict when and where earthquakes would occur. A wide range of nonscientific methods have been claimed to predict earthquakes: looking at the shapes of clouds or phases of the moon, peculiar animal behavior, pain in bunions or joints, and other crazier ideas. These people are usually ignored until, by chance, they manage to get one prediction right, and then they are suddenly swarmed by the press and given far more credibility than they deserve. Similarly, crackpots all over the Web make outlandish claims of success in predicting earthquakes. Without exception, however, these claims never hold up to scrutiny. In many cases, there is no clear evidence that the "prediction" was actually made in advance of the earthquake, but the crackpot has fraudulently claimed his prediction preceded the earthquake when it was actually made after. More often, the prognosticator made many predictions, all of which were wrong (but forgotten), and then gets one right by sheer accident. After all, there are earthquakes happening all over the world at any given time, so any random guess is right to some degree. This is a famous example of the human tendency for confirmation bias: believing that there is a causal connection between two coincidental events and ignoring all the other failed predictions. Like most "prophets," earthquake forecasters tend to make general, vague predictions, so they cannot be pinned down when a specific prediction fails and can claim that their vague prediction was on target when something happens. Therefore, there are actual organizations (such as the California Earthquake Prediction Evaluation Council) that examine pseudoscientists' claims and submit them to rigorous statistical and scientific testing. Not one prediction involving a specific date, time, place, and magnitude has held up to scrutiny.

The scientific attempts to predict earthquakes have gone through many phases of failure and success. The first concerted efforts began in the 1960s when seismologists attempted to find as many possible phenomena that might be precursors to an earthquake. Many different earth properties were examined, from the frequency of small earthquakes before a big shock, radon emissions from the ground, ground uplift and tilting, changes in the speed of P waves, electrical resistivity, fractoluminescence, satellite observations, and nearly everything else someone could think of. I vividly remember in graduate school during the 1970s when the hottest new theory, known as dilatancy, came along. Some of the professors in my graduate program at Lamont-Doherty Geological Observatory of Columbia University, such as Chris Scholz, were on the forefront of looking at dilatancy as an earthquake predictor. Dilatancy theory suggested that as rocks along a fault zone deformed they would expand, and this could be detected by changes in the groundwater, electrical conductivity, magnetic field, the tilt and uplift of the fault zone, and other phenomena. It seemed to be a successful theory when Chinese seismologists used it to predict the Haicheng earthquake of February 4, 1975. But the bubble was burst only a few months later when the Chinese failed to predict the Tangshan earthquake of July 27, 1976, resulting in the loss of more than a quarter of a million lives. In the 1990s, the Chinese government issued thirty false alarms but claimed to have successfully predicted the November 29, 1999, earthquake in Haicheng.

Other attempts at prediction have focused efforts on fault zones with a history of frequent earthquakes. The best known of these was the great experiment at Parkfield, California, in the California Coast Ranges halfway between Los Angeles and San Francisco. Since 1857, Parkfield has had an earthquake of at least magnitude 6 on average every 22 years (1881, 1901, 1922, 1934). The last such event was in 1966, so the U.S. Geological Survey and many other scientific organizations had hundreds of scientific instruments set up in the region by 1985, waiting for something to happen. The window of prediction was between 1986 and 1988, and those years passed uneventfully. By 1993, the 95% probability window had passed and still there was no earthquake. By 2000, most of the funding and staff had left, although many instruments were still in the ground running remotely from the U.S. Geological Survey in Menlo Park, California. Finally, on September 28, 2004, a magnitude 6 quake occurred—only sixteen years late. Valuable data were gathered by the waiting instruments, but seismologists were humbled by the failure of this "best bet" prediction.

The lesson from Parkfield and Tangshan is that no two earthquakes are alike. The precursors that successfully predict one event, like Haicheng in 1975, may not work for any other earthquake. Apparently, fault zones and earthquakes have a wide spectrum of behaviors, and so far, the Holy Grail of finding a universal short-term predictor has not been attained. Many seismologists believe it never will be, given the complexity of the problem of predicting earthquakes.

Another consequence of making earthquake predictions is liability. In this litigious American society, people will sue for the least little reason. If a scientist makes a firm short-term prediction and warns people to vacate a region, and then nothing happens, people will sue for the fear and inconvenience of having to flee for no reason. At least in the United States, this has held back short-term prediction more than anything else, whereas in China, the government seismologists can make one failed prediction after another, and the government orders evacuation after evacuation, but there is no way for the Chinese citizen to sue the government.

The ultimate absurdity occurred in June 2010, when the local prosecutors charged government seismologists with manslaughter for *not* predicting the April 6, 2009, earthquake near L'Aquila, Italy. As many people have pointed out, no seismologist should be making such predictions, for they are impossible and irresponsible. The entire prosecution is an absurd example of blaming the messenger and scapegoating for official incompetence. Science deals in probabilities and does not often make simple, clear predictions that people can follow. Seismologists are comfortable about giving probabilistic estimates for the likelihood of an event, but most do not feel that data are good enough (and may never be good enough) to make specific predictions that people can act on.

Consequently, most seismologists have focused on long-term predictions, giving a window of months to years when an event is likely to happen with a 95% probability. In this regard, they have been much more successful. The most common technique is the seismic gap method: identifying a stretch of an active fault zone where earthquakes are long overdue. This method was successfully used to predict the Loma Prieta earthquake a year before it struck in 1989, but the prediction was not specific enough for anyone to act on it. There have been many other similar successes in Alaska and elsewhere, but for all these successes, we also have "predictable" events like Parkfield that turned out to be unpredictable. In concert with this technique has been the growing technology of paleoseismology, where scientists dig trenches across fault zones and find evidence for ancient earthquakes that happened before historical records. These in turn provide some sense of how frequently and regularly a given stretch of fault has moved and the possibility of predicting the recurrence interval in that region. For example, Kerry Sieh of Caltech pioneered this field of research with his famous experiments at Pallett Creek, California, an ancient pond deposit just north of the Devil's Punchbowl and the San Gabriel Mountains. His trenches through the pond deposit uncovered traces of dozens of earthquakes over the past 2,000 years, all calibrated by carbon-14 dating of the peat deposits in the pond. He identified the trace of the 1857 Fort Tejon earthquake, the last "big one" in the region. Sieh predicted a recurrence interval of about 132 years, then about 147 years, making the prediction for the next big one between 1989 and 2004. As of 2011, it still had not happened and appears to be at least seven years overdue. However, the recurrence intervals vary from 45 to 332 years, so this predictor is not as precise as we would like.

The Psychology of Earthquakes

California tumbles into the sea
That'll be the day I go back to Annandale
—Steely Dan, "My Old School"

Of all natural disasters discussed in this book, earthquakes seem to strike the greatest fear and dread in people. The reasons are complex, but most psychologists agree that there are two frightening elements of earthquakes: unpredictability and loss of the sense of terra firma. Most other natural disasters, from hurricanes to tornadoes, to blizzards and even volcanoes, generally give some kind of warning, or at least can be tracked before they strike an area. No one can successfully predict earthquakes in the short term to aid evacuation, so earthquakes still strike without warning. People are also fearful that the solid ground beneath their feet is not so solid or that buildings can fall down on them without warning. Something deep and primal is disturbed when the most stable things in our lives, such as the earth beneath our feet, are no longer solid.

Hence, people fear earthquakes almost to the point of irrationality. As a

Southern California resident almost all my life, I've lived through every earthquake in this region. In 1971, I experienced the Sylmar quake. I woke up at dawn in our Glendale home and a lamp fell where my head had just come off the pillow as I sat up awakened by shaking. I just missed a concussion or being killed. I also experienced the 1987 Whittier quake, which woke me up early one morning before my Introduction to Geology class. I set aside my planned lecture and gave an impromptu lecture on earthquakes (supplemented by what I'd heard on the morning news reports) to a packed lecture hall—not just the students enrolled in the class, but many others who wanted to hear about the earthquake. On January 17, 1994, I was awakened at 4:30 a.m. when the Northridge quake struck. Our house in Sunland shook badly, but only a few items were broken. I had to crawl in the darkness across the bed covered in glass shards from a fallen mirror to find my crying two-year-old son, pull him out of the broken glass, and then comfort him until he went back to sleep. For him, the biggest inconvenience was the loss of power, so we couldn't turn on the lights or the TV or play a video to calm and reassure him.

Having experienced these large earthquakes, and hundreds of smaller ones, I've learned to cope with them mentally. I keep preparations for an earthquake emergency in a safe place. When the big one strikes, everyone in my family knows what to do, and we have food, water, and supplies ready to keep us going for days. I am constantly questioned by people who have never experienced an earthquake and are frightened at the prospect, even though they live in areas where deadly tornadoes, hurricanes, or blizzards occur every year. As a graduate student teaching assistant at Columbia University in New York, I remember a Barnard professor scared the dickens out of the big "rocks for jocks" Introduction to Geology class with exaggerated stories about earthquakes, which he'd never actually experienced—while I, veteran of many quakes, found his fear laughable.

Many urban myths have sprung up regarding earthquakes, probably because of how people fear them. It has been mentioned how earthquakes do *not* form giant chasms filled with lava, as seen in the first Christopher Reeve *Superman* movie—only long straight valleys with features like offset roads and fence lines (fig. 1.3B). Chasms open up when landslides are triggered by earthquakes, but these are not on the fault line (plate 1). Another urban legend is expressed in the lyrics of the Steely Dan song, "My Old School": "California tumbles into the sea." In actuality, California west of the San Andreas fault is sliding northward a few meters each time there is a major earthquake, and in 50 million years, California will be next to Alaska, not tumbling into the sea.

Many people believe in the myth of "earthquake weather." Supposedly, great earthquakes happen during unusually hot days. In reality, there is no correlation between the occurrence of earthquakes and weather, daytime temperature, or time of day. Earthquakes happen around the clock every few minutes during any time of day and any kind of weather. For example, the three big earthquakes I've experienced all happened early in the morning during the relatively cool winter and spring months. If you think about it, the lack of real "earthquake weather" makes sense. The daily temperature fluctuations we experience in the air only penetrate down a few inches into the ground. Dig a few inches down in the dirt on a hot day—it's quite cool below the surface. Yet earthquakes are generated kilometers down in the crust and never experience any daily temperature fluctuations. This myth is a product of selective memory. People experience a traumatic event and happen to notice that it occurred on a hot day, hence the legend of earthquake weather. Look at the entire record of earthquakes over time and the correlation vanishes.

If you adopt the viewpoint of an insurance adjuster, an actuary, or a statistician, there is no real connection between our exaggerated fear of earthquakes and actual risk. People fear earthquakes far more than they fear tornadoes, hurricanes, blizzards, lightning, or driving their cars, yet every one of these phenomena kills far more people per year or per decade than earthquakes do. Let's look at some numbers.

In the 104 years since the 1906 San Francisco quake, fewer than 600 Americans have lost their lives in an earthquake. (The numbers are far higher in the less-developed parts of the world, because buildings are constructed with unreinforced masonry and other cheap but dangerous construction that is a deathtrap in an earthquake.) By contrast, tornadoes kill dozens to hundreds of

54 Catastrophes!

Americans each year, as do floods and blizzards (see chapter 12). Hurricanes kill even more, even though people are warned to evacuate days in advance. Heat waves kill thousands each year, even more now that climate change has ratcheted up the global thermostat. The chances of being struck by lightning are higher than the chances of being killed in an earthquake. Even driving a car is more dangerous than an earthquake, yet people are not deathly afraid of their cars, and many drivers are careless or reckless. In 2007, 41,000 people in the United States died in traffic accidents! By any measure, tornadoes, hurricanes, blizzards, floods, lightning, and cars are hundreds to thousands of times more deadly than earthquakes, yet only earthquakes generate a special fear and terror.

Thus, there is a fundamental disconnect between a rational assessment of real risk and our irrational fear of the unknown. Our smart brains and science and statistics tell us that earthquakes should not be so terrifying, but our primal instincts and lack of education about real risks always seem to trump reality.

FOR FURTHER READING

Bolt, B. A. 2004. *Earthquakes*. 5th ed. W. H. Freeman, New York.
Clarke, T. 1996. *California Fault: Searching for the Spirit of a State along the San Andreas*. Ballatine Books, New York.
Lawson, A. C., ed. 1908. The California earthquake of April 18, 1906. Report of the State Earthquake Investigation Commission. *Carnegie Institution of Washington Publication* 87, Washington, DC.
Prentice, C. S. 1999. San Andreas fault: the 1906 earthquake and subsequent evolution of ideas. *Geological Society of America Special Paper* 338:79–85.
Shrady, N. 2008. *The Last Day: Wrath, Ruin, and Reason in the Great Lisbon Earthquake of 1755*. Viking, New York.
Ulin, D. L. 2004. *The Myth of Solid Ground: Earthquakes, Prediction, and the Fault Line between Reason and Faith*. Penguin Books, New York.
Walker, B. 1982. *Earthquake*. Time-Life Books, Chicago.
Winchester, S. 2006. *A Crack in the Edge of the World: America and the Great California Earthquake of 1906*. Harper, New York.
Zeilinga de Boer, J., and D. T. Sanders. 2005. *Earthquakes in Human History*. Princeton University Press, Princeton, NJ.

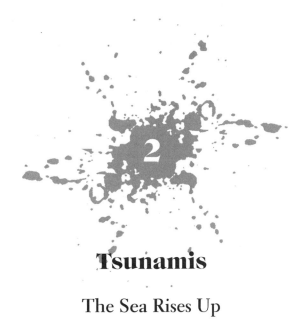

Tsunamis

The Sea Rises Up

Within minutes, the beach and the area behind it had become an inland sea, rushing over the road and pouring into the flimsy houses on the other side. The speed with which it all happened seemed like a scene from the Bible—a natural phenomenon unlike anything I had experienced before. As the waters rose at an incredible rate, I half expected to catch sight of Noah's Ark.
—Michael Dobbs, *Washington Post*, December 2004

The Indian Ocean Tsunami: Boxing Day, December 26, 2004

The southern regions of Thailand and the north coast of Sumatra are heavily populated, and during the Christmas holidays, they are crowded with tourists seeking refuge from Northern Hemisphere winters as they enjoy the sun and pristine beaches. The day after Christmas (Boxing Day on the British calendar) a violent earthquake struck 160 km (100 miles) off the northwest coast of Sumatra. Its moment magnitude was 9.2, the second-largest earthquake ever recorded on a seismograph, and lasted between 8 and 10 minutes, the longest duration of shaking ever observed. The entire place vibrated by as much as 1 cm (0.5 inches) up and down and triggered sympathetic earthquakes as far away as Alaska. As the Indian plate is pushed under the Burma plate, it produces a huge subduction zone that is responsible for the island nations of Indonesia and Malaysia. The fault line of this plate boundary formed a rupture about 400 km (250 miles) long and 100 km (60 miles) wide, which was located 30 km (19 miles) beneath the seabed—the longest rupture ever caused by an earthquake. The energy released by the quake was about 550 million times more powerful than the atomic bomb that destroyed Hiroshima.

Although the earthquake shook Thailand, Indonesia, and Malaysia that morning, many people recovered quickly, or ignored it, and went on with their day. Minutes after the earthquake, however, the effects of the submarine quake showed up on the western shore of Thailand and Indonesia. First, the sea level dropped abnormally low, exposing the coastal reefs and seabed. People ran out for the easy fishing, only to discover to their horror that the unusually low sea level was a trap; suddenly a 30-m-tall (100-foot-tall) seismic sea wave (fig. 2.1), or tsunami, bore down on them.

The largest loss of life occurred along the western coast of Indonesia, Malaysia, and Thailand. Waves propagated westward across the Indian Ocean at 500 to 1,000 km/h (310 to 620 mph). Within 90 minutes, waves had reached Sri Lanka and the coast of India. Seven hours later the tsunami reached Africa's Somalia coast, where the damage and loss of life were worse than in Bangladesh, which was much closer to the source of the waves (but not in the direct path of the main wave propagation). The tsunami was noticed as far as South Africa's western coast, 8,500 km (5,300 miles) away, and propagated out into the Pacific Ocean, although with less force.

The evening of the tsunami Michael Dobbs of the *Washington Post* (2004) reported this account of the tsunami as it reached Sri Lanka:

Disaster struck with no warning out of a faultlessly clear blue sky. I was taking my morning swim around the island . . . when I heard my brother shouting at me, "Come back! Come back! There's something strange happening with the

Fig. 2.1. People beginning to flee the crashing wall of water from the 2004 Indian Ocean tsunami. (Image courtesy Wikimedia Commons)

sea" . . . As I swam to shore, my mind was momentarily befuddled by two conflicting impressions: the idyllic blue sky and the rapidly rising waters. In less than a minute, the water level had risen at least 15 feet—but the sea itself remained calm, barely a wave in sight . . .

There are reports of hundreds, perhaps thousands of people missing and drowned in southern Sri Lanka. The coastal road is littered with carcasses of boats and dead dogs. Even a few dead sharks have washed up on the road. We have no water, and no electricity and are practically cut off from the rest of Sri Lanka . . . The holiday that we planned and dreamed about for many months is in ruins. We feel fortunate—fortunate to be alive.

News of the tsunami spread around the world. Tourists' video footage showed horrifying pictures of a wall of water rushing through villages and crushing and sweeping away everything in its path (fig. 2.1). News media broadcast satellite images of coastal devastation (figs. 2.2, 2.3), as well as reports

January 1, 2004

December 26, 2004

Fig. 2.2. Satellite images of the Kalutara, Sumatra, coast before the 2004 tsunami (*top*) and during the tsunami (*bottom*). Note how much of the coast region was simply swept away by the surge and backwash of the huge waves, which still cover the land in the bottom photo. (Photos courtesy NASA Images)

May 18, 2004

erosion — — erosion

— destroyed pier

January 7, 2005

Fig. 2.3. Satellite image of Meulaboh, Indonesia, showing the coastal features that were destroyed by the 2004 tsunami and the complete destruction of the village on this peninsula as the tsunami waves swept across it. (Photos courtesy NASA Images)

of damage from other parts of the world (plate 4). The staggering death toll included more than 167,000 in Indonesia, 35,000 in Sri Lanka, 18,000 in India, and 8,000 in Thailand; a dozen Asian and African countries on the Indian Ocean also suffered. As many as 543 Swedish tourists in Thailand and Indonesia were killed. Overall, an estimated 250,000 people died, and many millions more were displaced, making this the worst natural disaster in recorded history and the deadliest tsunami ever recorded.

The disaster could have been worse, but many Indonesians, familiar with earthquakes and tsunamis, had retreated to higher ground after the quake ended. Many tourists and locals, however, were less informed or were seduced by the unexpected bounty of fish flopping around for easy collection as the sea withdrew. On a beach in Phuket, Thailand, a 10-year-old British tourist had studied tsunamis in her geography class and recognized that the receding ocean and frothing bubbles meant disaster. She warned her parents, and the entire beach was evacuated, resulting in no loss of life. A biology teacher from Scotland recognized the signs at Kamala Bay north of Phuket and led a busload of vacationers and locals to higher ground.

What Is a Tsunami?

The 2004 Indian Ocean tsunami was so catastrophic that the world soon learned much about this relatively rare phenomenon. *Tsunami* is a Japanese word that means "harbor wave." The Japanese have experienced many tsunamis resulting from frequent offshore earthquakes. One of the most devastating tsunamis occurred on June 15, 1896. Beaches were crowded with Japanese tourists, and a fishing fleet was at sea when an offshore earthquake rocked the seafloor. About 20 minutes later, the sea rapidly retreated, followed 45 minutes later by a huge wave 29 m (95 feet) high. It crashed through the coast of southwestern Japan, destroying more than 10,000 homes and killing more than 27,000 people. Fishermen in the open ocean barely noticed the large swells as the waves swept under them. They returned home to coastal devastation, their villages gone and their families dead.

A tsunami is a seismic sea wave caused by a sudden displacement of water

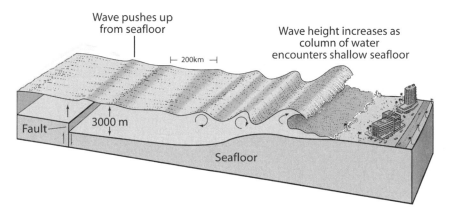

Fig. 2.4. Diagram of tsunami wave motion. (Diagram courtesy USGS, redrawn by Pat Linse)

in the ocean, as during an earthquake, a submarine landslide, or a volcanic eruption. It is not a tidal wave. A tsunami has nothing to do with the tides or with the waves formed by storm winds. A positive outcome of the 2004 Indian Ocean event is that people worldwide learned this distinction, and, after decades of misuse, the media at last stopped describing tsunamis as tidal waves.

When large earthquakes, submarine landslides, or volcanoes displace the seafloor (fig. 2.4), they generate a pulse of energy up and away from this displacement that forms waves propagating outward in all directions like ripples on a pond. These waves have long wavelengths and low amplitudes, so they only cause the sea to swell up a few meters in the open ocean and are usually difficult to detect by boats at sea. However, as waves approach the shallows near land, the sea bottom disrupts the motion of the wave, which has a deep wave base. The swells pile up near the coast and rise out of the ocean, forming enormous waves much larger than those produced by any storm or surf disturbance. In contrast to the Hollywood stereotype of a beautifully curved wall of water rising higher than buildings, a real tsunami looks like any ordinary foamy, crashing, breaking wave but much larger. The biggest difference between a tsunami and a normal sea wave is that the energy of a tsunami does not end in the normal surf zone. Instead, water rushes far inland, sweeping through towns and leveling buildings with its enormous power.

The April's Fool Tsunami: Alaska and Hawaii, 1946

Although the most recent tsunami occurred in the Indian Ocean, most deadly tsunamis occur in the Pacific Ocean. The Pacific Rim is known as the Ring of Fire (fig. 1.6) because it is surrounded by subduction zones and a few transform faults, which are all tectonically active. Many of the world's earthquakes and volcanoes occur there, generating opportunities for tsunamis. The Pacific has few continents or islands in the way; therefore, a large earthquake on one edge can propagate tsunamis across the rest of the Pacific within hours (fig. 2.5).

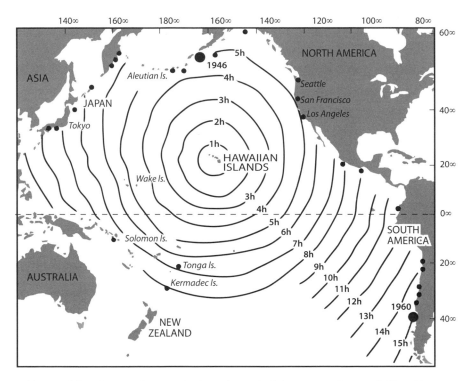

Fig. 2.5. Map showing the time (in hours) required for a tsunami to reach Hawaii from the edge of the Pacific. (Diagram courtesy USGS, redrawn by Pat Linse)

Fig. 2.6. Before and after photos of the destruction of the Scotch Cap lighthouse in the Aleutians due to the April 1, 1946, tsunami. (Photos courtesy of USGS Photo Library)

The Aleutian earthquake and tsunami of April 1, 1946, is a classic example of a tsunami's power. Two huge subduction movements generated large earthquakes (magnitude 7.8), which immediately triggered tsunamis around the Aleutians. In the Scotch Cap lighthouse on Unimak Island (fig. 2.6), five workers were shaken awake and wondered what the dark night might bring. However, they felt safe because the lighthouse had been built to withstand the strongest Pacific storms, with its steel-reinforced concrete walls and a base 14 m (46 feet) above sea level. The highest parts of the five-story-tall lighthouse appeared to be safely above the worst storm waves, but tsunamis are not ordinary storm waves. Twenty minutes after the earthquakes, a wall of water 30 m (100 feet) high pounded and destroyed the lighthouse, killing the five men (fig. 2.6). All that remained were piles of twisted steel, broken concrete, and the foundation.

From there, the tsunami moved quickly across the Pacific, first traveling about 780 km/h (485 mph) and then slowing to 56 km/h (35 mph) as it neared the north shore of Hawaii (figs. 2.5, 2.7) five hours after the initial earthquake. Waves moving at 35 mph are faster than any human can run, so the people along the wharf had no chance and drowned immediately. Farther inland inhabitants were also surprised when the foamy wave crashed through buildings, carrying boats and destroying most of the buildings of downtown Hilo. Waves moved inland for almost a kilometer in many places, wiping out nearly the entire coastal region of the north coast of Hawaii.

One eyewitness was police officer Robert "Steamy" Chow:

"April Fool! April Fool!" That's what 24-year old Hilo police officer Robert "Steamy" Chow replied as he drove to work when a poi shop owner shouted to him that a tsunami was coming. But of course, it was no joke. It was early morning, April 1, 1946 and Hilo was about to be changed forever. Officer Chow had joined the police force only about three years earlier, when the department was looking for Cantonese-speaking recruits to communicate with the many Chinese plantation workers . . . When he was on Wailuku Drive and Keawe Street, the next wave hit. It took the railroad bridge, and water went underneath his car. People were running in front of his car. Officer Chow remained on duty for the next 20 hours or more. He remembers how he had to go to the electric company's locker that was being used as a morgue. (Pacific Tsunami Museum, www.tsunami.org/)

The Great Chile Quake and the 1960 Hilo Tsunami

Hawaii is in the center of the bull's-eye of the Pacific. It has been hit by more tsunamis than any other place on earth (fig. 2.5). On Sunday, May 22, 1960, the most powerful earthquake ever recorded struck southern Chile at 3:11 p.m., generating a huge tsunami. Most Chileans, already familiar with earthquakes and tsunamis, headed for higher ground after the earthquake ended. Fifteen minutes after the shaking stopped, the sea rose rapidly, cresting at 4.5 m (15 feet) above sea level. With incredible speed, the wave surged back to the sea, dragging broken houses and boats with it, making loud hissing and gurgling noises.

Fig. 2.7. A, Tsunami breaking over Pier 1 in Hilo, Hawaii, on April 1, 1946. The man in the left foreground was one of the 159 killed. B, The Army Crash Boat, which was docked before the tsunami, washed onshore about 400 feet, over the railroad tracks and up against these molasses tanks. C, People fleeing the wave, visible on the right next to the man's head. (Photos courtesy NOAA)

Some Chilean boat owners thought that this smooth outgoing wave could be ridden safely out to sea, and tried to save their boats that way. But at 4:20 p.m., a second tsunami arrived, traveling at 200 km/h (125 mph), cresting at 8 m (26 feet). The wave crushed boats pushed shoreward, drowned their passengers, and destroyed remaining buildings on the waterfront left standing after the earthquake and the first tsunami. A third wave—larger than the first two— followed, cresting at 11 m (35 feet). By the time the tsunamis had stopped, more than 1,000 Chileans were dead.

The energy from this monster quake moved across the Pacific toward Hawaii. After the 1946 Hilo tsunami, a warning system had been established. Sirens sounded and the western shore of Hawaii's Big Island as well as Hilo were evacuated. Most people had evacuated before the first wave hit at 9:57 p.m., one minute after its predicted arrival time. Nevertheless, some sightseers went to the shore to witness it and were promptly drowned by the monstrous wave (fig. 2.8A). Those sightseers constituted most of the 61 people killed. Many more would have died without the early warning system. Still, the streets of Hilo were again devastated by the huge wall of water (fig. 2.8B).

Officer Robert "Steamy" Chow was with the Hilo police force 14 years after the 1946 Hilo tsunami and witnessed the effects of the 1960 tsunami:

> Just minutes after 1 am, Officer Chow was driving on Keawe Street when the whole night sky filled with a blinding blue flash and the town became dark. The power plant had exploded when slammed by the powerful third wave. Now the chaos began—people fleeing, whole city blocks turned to splinters in the dark night . . . It wasn't until the next morning that [Chow's] wife found him . . . Steamy Chow is still one of Hilo's most beloved citizens. (Pacific Tsunami Museum, www.tsunami.org/)

The 1960 Chilean earthquake tsunami continued to wreak havoc. It raced on to Japan, where another 185 people were killed, despite warnings of its impending arrival. The earthquake's energy pulse bounced back and forth across

Fig. 2.8. A, People foolishly standing outside the Suisan Fish Market, waiting to see the tsunami arrive. They were all killed. B, Devastation of the streets of Hilo after the 1960 tsunami. (Photo courtesy of NOAA)

the Pacific basin like echoes that slowly die out. Tidal gauges were still recording its power around the Pacific a week after the event.

Because of the long history of tsunamis on the Pacific, most coastal regions have tsunami warning systems, and the chances of loss of life are diminished. However, the Indian Ocean had few recorded tsunamis in historic memory, so no such system was installed, resulting in greater loss of life. An effort is under way to develop a warning system for the Indian Ocean, but it is years away and too late for the 230,000 people who died in the 2004 tsunami.

Landslide-Induced Tsunamis: Papua New Guinea, 1998

Although most tsunamis are due to seismic activity, other ocean disturbances can produce huge sea waves. At 6:49 p.m. on July 17, 1998, a magnitude 7.1 earthquake occurred 25 km (16 miles) off the north coast of Papua New Guinea. The earthquake was large and caused a vast submarine landslide off the continental shelf of New Guinea, which displaced a huge volume of sediment nearly 5 km (3 miles) from the source. This massive gravity slide greatly amplified the volume of water displaced upward into seismic sea waves. Because the event was so close to the coastline, the tsunami arrived almost immediately after the earthquake, hitting the coastline and shooting spray up to 30 m (100 feet) high. Witnesses compared the sound to distant thunder as the water swept in, and then the sea pulled back far beyond normal low tide. After another 4–5 minutes, the rumbling returned, and a wave about 4 m (13 feet) high approached. Most residents close to the shore could not outrun the wave and were swept 200 m (660 feet) inland or out to sea as the wave washed back to the ocean. A few minutes later, another wave, 14 m (45 feet) tall, appeared, which poured across the coastal regions at speeds up to 24 km/h (15 mph). A three-wave sequence then washed hundreds of people out of their houses and out to sea or into the lagoon behind the barrier island. The island was swept clean of its four villages (fig. 2.9), with a loss of at least 2,200 people—mostly children, who were unable to swim or run fast enough to escape. In some places, the death estimates were much higher (3,000–4,000), but an accurate

Fig. 2.9. Effects of the 1998 tsunami in Papua New Guinea. A two-story wooden school building that stood near the church at Sissano Mission was carried 65 m by the wave until caught by a grove of coconut palms. The lower floor of the building collapsed, but the upper floor classrooms were preserved. Schoolwork was still hanging on the wall. (Photo courtesy NOAA)

count was impossible to make because many people were washed out to sea, and bodies were scattered across the landscape. Instead of recovering bodies, the government of Papua New Guinea closed off the area and allowed the bodies to decompose naturally. At least another 10,000 people were displaced and became refugees.

Survivors suffered even more. The international community tried to provide aid, but the mostly remote region was hard to reach, and roads and communications had been wiped out. Bodies of loved ones were left to rot while survivors scrounged for food and water and tried to eke out a living. To limit the spread of disease, the government sent gunmen in to kill animals feasting on the dead bodies. Health care was minimal. Many badly injured people suffered from broken limbs. As a result, many people became infected before they could be treated and had to have limbs amputated because of gangrene. The effects of the 1998 tsunami remain today, a full decade later. The region is still desperately poor and struggling.

Volcanic Tsunamis: Krakatau 1883

Huge submarine volcanoes that explode and displace water during the eruption can also trigger tsunami waves as powerful as those created by an earthquake. The most famous are the tsunamis that resulted from the eruption of Krakatau in 1883. The eruption will be discussed further in chapter 3, but the tsunami activity deserves special mention.

The eruption took place in several phases, but the enormous fourth eruption that blew the top off the mountain caused the biggest tsunami. The eruption caused the northern two-thirds of the island to vanish, sending shock waves in all directions, followed by the collapse of the caldera, which also displaced water as it poured in to fill the gap. Within an hour after the explosion, huge waves up to 22 m (72 feet) high inundated nearby coastal regions. A wave 46 m (150 feet) high, the largest tsunami ever recorded, destroyed the town of Merak. Waves leaving the Sunda Strait propagated up and down the coasts of Java and Sumatra, where surges as high as 15 m (45 feet) wiped out most of the villages. At least 30,000 people and maybe as many as 126,000 are thought to have died, far more than were killed directly by the Krakatau eruption. The precise number of dead can never be known. Bodies of victims were found far out to sea weeks after the event, because they had been washed so far from shore and drowned.

Powerful waves picked up huge coral blocks weighing 600 tons and carried them inland. The steamship *Berouw* (fig. 2.10) was carried 3 km (2 miles) inland by the incoming wave and stranded atop trees 10 m (33 feet) above sea level; its 28 crew members perished. Few people survived, and only a handful of eyewitness accounts are known, including this account recorded in the town of Anjer in Java:

> At first sight, it seemed like a low range of hills rising out of the water, but I knew there was nothing of the kind in that part of the Sunda Strait. A second glance — and a very hurried one it was — convinced me that it was a lofty ridge of water many feet high . . . There was no time to give any warning, and so I

Fig. 2.10. Contemporary engravings of the Royal Dutch Navy's armed paddle steamer *Berouw* being hit by the tsunami wave (A), and then carried inland and stranded far up the river by the Krakatau tsunami (B). (From Wikimedia Commons)

turned and ran for my life. My running days have long gone by, but you may be sure that I did my best. In a few minutes, I heard the water with a loud roar break upon the shore. Everything was engulfed. Another glance around showed the houses being swept away, and the trees thrown down on every side. Breathless and exhausted I still pressed on . . . A few yards more brought me to some rising ground, and here the torrent of water overtook me. I gave up all for lost . . . I was soon taken off my feet and borne inland by the force of the resistless mass. I remember nothing more until a violent blow revived me . . . I found myself clinging to a coconut palm. Most of the trees near the town were uprooted and thrown down for miles, but this one fortunately had escaped and myself with it . . . As I clung to the palm-tree, wet and exhausted, there floated past the dead bodies of many a friend and neighbor. Only a mere handful of the population escaped. (Officer and Page 1993, 17–18)

Twelve hours later, the tsunamis from Krakatau reached the Gulf of Aden on the Arabian Peninsula, having traveled about 300 mph across the Indian Ocean. Ships in South Africa were rocked violently as they lay in harbor. Smaller oscillations were reported in Japan, Australia, Hawaii, Alaska, and California, as the waves worked their way out of the Indonesian archipelago and across the Pacific. Even tidal gauges in the English Channel recorded a disturbance.

Tsunami Clues

Even though tsunamis happen rarely during a single human lifetime, they are important geologic forces in tsunami-prone regions. Just as paleoseismologists dig trenches through fault zones to reveal a region's history of earthquakes, geologists can look for buried evidence of tsunamis. Records of ancient tsunamis can extend the historical and instrumental records to the recent past and fill in the gaps of historic and geographic accounts. An international team led by Kruawun Jankaew of Chulalongkorn University in Thailand dug hundreds of pits in Thailand and Sumatra to reveal this record. A typical pit shows an alternation of thick, light-colored sand layers interrupting the normal dark soil. The top sand layer was deposited by the 2004 tsunami; successive lower layers of sand indicate older events over the last few thousand years. By carbon-14 dating bark fragments in soil below the second sand layer, the next oldest event before the 2004 tsunami occurred between AD 1300 and 1450. In addition, two earlier tsunamis during the past 2,500 to 2,800 years are evident.

Geologists who rushed to Papua New Guinea after the 1998 tsunami undertook similar research. They mapped the sand sheets washed inland, dug trenches for evidence of ancient tsunamis, and mapped the direction of waves and devastation from fallen trees. The scientific study of modern and ancient tsunamis has grown remarkably because of the events of 1998 and 2004. Now, at any geological meeting, several talks and even an entire session are held on ancient tsunamis. In the past 20 years, there have been similar studies of tsunami deposits from Japan, Kamchatka, Alaska, Cascadia, and Chile. A major workshop on tsunami deposits, sponsored by the National Science Foundation, was held at the University of Washington in 2005, and the subsequent report forms the foundation for ongoing research on ancient tsunamis. How to distinguish a tsunami deposit from any other large storm deposit is the biggest issue geologists confront. Geologists look for evidence of tidal deposits that show sudden subsidence or uplift along a coastline, which would indicate a past earthquake not produced by an ordinary storm. In addition, the association of the suspected tsunami deposits with features due to liquefaction of the sediment by the shaking of an earthquake (similar to the way in which quicksand responds to a disturbance) is considered strong evidence that a deposit is tsunami related and not just a storm product. Sudden changes in sea level, such as the depth preferences of the marine life preserved in sediment, also help to diagnose tsunami sequences. In some cases, the geochemistry of the sedimentary particles and fossils might help distinguish different deposits because hurricanes and other storms are formed largely of fresh water, whereas tsunamis are entirely from marine water.

Tsunamis and the Death of the Dinosaurs?

Perhaps the most controversial example of a potential ancient tsunami deposit is the sequence of rocks found at the end of the age of the dinosaurs. According to the prevailing interpretation, tsunamis that emanated from the site of the meteorite impact on northern Yucatán formed this sequence. The impact supposedly wiped out the dinosaurs and more than 50 percent of the marine species on the planet. The rock sequence at the boundary event is best exposed in the Brazos River region on the Texas Gulf Coastal Plain and has been repeatedly studied by scientists who think it is an undoubted impact-derived tsunami (Bourgeois et al. 1988) and by others who argue that this sequence is a large storm deposit spanning a much longer interval of time (Keller and Stinnesbeck 1996). A layer rich in droplets of crustal material ejected from the impact crater ("spherules") is found, followed by a series of thick, sandy layers interpreted as tsunami deposits by one school of thought or simple storm deposits by opponents. Above this is a burrowed sandy layer, which impact advocates claim is the final phase of the tsunami, but impact skeptics argue is a normal marine deposit on top of the sandy layers. In their view, this deposit shows that a long time elapsed between the impact and the final extinction of dinosaurs and late Cretaceous marine plankton (which are supposedly found above the tsunami layers).

Like everything concerning the dinosaurs' extinction (see chapter 11 and Prothero [2006], chapter 2; Prothero [2009], chapter 5), the events of the end of the Cretaceous are complex and do not lend themselves to the easy explanation that a single impact caused their extinction. There is little doubt that an extraterrestrial impact occurred on northern Yucatán near the end of the Cretaceous and that it spread deposits all around the Caribbean and the Gulf of Mexico. Abundant evidence also indicates that life was undergoing tremendous stresses and extinction well before the impact took place, and many animals (crocodiles, turtles, amphibians, most insects) could not have survived a world as hellish as impact advocates describe. Whether the Brazos River sequence proves that life returned to normal well after the storm beds and before the mass extinction is still open to debate. Whatever the conclusion, however, this argument has demonstrated the importance of recognizing ancient tsunami deposits and learning the lessons from the past that they offer.

FOR FURTHER READING

Blackhall, S. 2005. *Tsunami*. Taj Books, New York.

Bourgeois, J., T. A. Hansen, P. L. Wiberg, and E. G. Kauffman. 1988. A tsunami deposit at the Cretaceous-Tertiary boundary in Texas. *Science* 241:567–70.

Funk, J. 2005. *Tsunami*. Triumph Books, New York.

Keller, G., and W. Stinnesbeck. 1996. Sea level changes, clastic deposits and megatsunamis across the Cretaceous-Tertiary boundary. *In* N. MacLeod and G. Keller, eds., *The Cretaceous-Tertiary mass extinction: Biotic and environmental events*, 415–50. W. W. Norton, New York.

Powers, D. M. 2005. *The Raging Sea: A Powerful Account of the Worst Tsunami in U.S. History*. Citadel, New York.

Rhodes, B., M. Tuttle, B. P. Horton, L. Doner, H. Kelsey, A. Nelson, and M. Cisternas. 2006. Paleotsunami research. *Eos, Transactions, American Geophysical Union* 87 (21): 205–9.

Tibballs, G. 2005. *Tsunami: The Most Terrifying Disaster*. Carlton Books, New York.

Volcanoes

Hell's Cauldron

He also had one volcano that was extinct. But, as he said, "One never knows!"
So he cleaned out the extinct volcano, too. If they are well cleaned out,
volcanoes burn slowly and steadily, without any eruptions. Volcanic eruptions
are like fires in a chimney. On our earth we are much too small to clean out
our volcanoes. That is why they bring no end of trouble upon us.
—Antoine de Sainte-Exupéry, *The Little Prince*

The Wrath of Vulcan: Vesuvius, AD 79

The year AD 79 was important in Roman history. A decade earlier, the Roman
general Vespasianus had taken over as emperor, ending a year of civil war and
anarchy following the death of Nero. Vespasianus had stabilized the empire,
gotten the imperial accounts back on the profitable side, and improved the
political situation for the middle and lower classes. He began many important
construction projects, including great temples and the mighty Colosseum,
completed in AD 79, built on the site of one of Nero's palaces. At his death on
June 24, his son Titus succeeded him. Titus proved to be an even more com-

petent emperor, expanding the empire in Wales and Scotland and suppressing revolts in Palestine.

For the residents of towns around the Bay of Naples, AD 79 was also quite eventful. The Naples area was then, as it is now, a sleepy port city popular with vacationers, fishermen, and boaters. Since the reign of second emperor Tiberius, the Roman rulers had had a private villa on the island of Capri, on the southern edge of Naples Bay. Tiberius left Rome altogether in his final years and ruled the empire from his villa. By AD 79, the bay was crowded with towns, and the outlying areas were famous for their agriculture, especially grape and wine growing. Indeed, there were vineyards all around the slopes of Mt. Vesuvius, the great volcano to the northeast.

The Romans probably did not realize that Vesuvius was an active volcano, because its last eruption was more than two centuries before in 217 BC. For the 17 years since the great earthquake of AD 62 that had destroyed much of Pompeii, Herculaneum, and Neapolis (Naples), there had been frequent earthquakes in the region. As early as 30 BC, the Greek historian Diodorus Siculus described the Campanian plain as "fiery" (*Phlegrean*) because Vesuvius showed signs of the fires that had burned long ago. The Romans thought that the fires of Etna were due to the forges of Vulcan (Hephaistos to the Greeks), the god of the fire. He used the heat of the underworld to hammer out armor, metalwork, and weapons for the gods (including the thunderbolts thrown by Zeus/Jupiter). When eruptions occurred, it was said that Vulcan was angry because his wife Venus had cheated on him. The Romans considered Vesuvius sacred to Hercules (Herakles to the Greeks), and some scholars believe that the name "Vesuvius" is derived from the Greek for "son of Zeus" (as Herakles was Zeus's son). The Romans named the port city of Herculaneum in honor of Hercules. Despite all these warnings, however, the area around the mountain was heavily populated, with 20,000 people in the town of Pompeii alone, and farming towns all around the base of Vesuvius. Then, as now, the volcanic soil was too rich and the climate too good for people to fear the long-dormant volcano. Even today, the area around Vesuvius has the highest population density of any active volcano in the world.

Fig. 3.1. Eruption of Mt. Vesuvius in 1944. (Photo courtesy USGS Photo Library)

By early August, the earthquakes were more frequent, but most residents ignored them. In addition, springs and wells had dried up, suggesting that the water table was dropping. On August 23, the Romans had celebrated the festival of Vulcanalia, to honor the god Vulcan. Then early in the warm summer's afternoon of the following day, August 24, Vulcan gave his reply. First was a huge explosion, and the sky darkened as ash and pumice rained down on inhabitants for 18–20 hours (fig. 3.1). Some residents of Pompeii and Herculaneum evacuated immediately, but many remained behind who were unwilling or unable to leave because there were not enough boats in the harbor to carry them, and the roads were blocked by volcanic activity or clogged with traffic. After this phase of eruption, the streets of Pompeii were filled with 2.8 m (9 feet) of ash and pumice, making it harder to evacuate, let alone breathe.

Then after almost a day of this rain of hell, the mountain let loose another type of eruption: many *nuées ardentes* (glowing clouds), or pyroclastic flows, a superheated (up to 850°C, or 1,560°F) mixture of volcanic gases and ash that roared down the mountain slope at 160 km/h (100 mph), incinerating everything in its path, and burying Herculaneum under tens of meters of volcanic deposits, or tuff.

Most eyewitnesses did not record their experiences, or their accounts have been lost in the mists of history. Fortunately, we do have one excellent eyewitness account, written by the historian Pliny the Younger. He was 17 at the time and fleeing with his family in a boat to the town of Misenum, across the bay 35 km (22 miles) from the volcano. In a letter to his friend, the historian Cornelius Tacitus, the younger man described how his 56-year-old uncle, Pliny the Elder, one of Rome's leading admirals, scholars, and naturalists, decided to take a boat closer to the mountain to rescue his friends. It is one of my favorite accounts of any eruption because I first read it in the original in my high school Latin class:

My dear Tacitus,

You ask me to write you something about the death of my uncle so that the account you transmit to posterity is as reliable as possible. I am grateful to you, for I see that his death will be remembered forever if you treat it [in your Histories]. He perished in a devastation of the loveliest of lands, in a memorable disaster shared by peoples and cities, but this will be a kind of eternal life for him . . .

He was at Misenum in his capacity as commander of the fleet on the 24th of August [AD 79], when between 2 and 3 in the afternoon my mother drew his attention to a cloud of unusual size and appearance. He had had a sunbath, then a cold bath, and was reclining after dinner with his books. He called for his shoes and climbed up to where he could get the best view of the phenomenon. The cloud was rising from a mountain—at such a distance we couldn't tell which, but afterwards learned that it was Vesuvius. I can best describe its shape by likening it to a pine tree [today, we would compare it to a "mushroom cloud"]. It rose into the sky on a very long "trunk" from which spread some "branches." I imagine it had been raised by a sudden blast, which then weakened, leaving the cloud unsupported so that its own weight caused it to spread sideways. Some of the cloud was white, in other parts there were dark patches of dirt and ash. The sight of it made the scientist in my uncle determined to see it from closer at hand. [This style of explosive mushroom cloud of ash and pumice is now called a "Plinian eruption" in his honor.]

He ordered a boat made ready . . . The expedition that started out as a quest for knowledge now called for courage. He launched the quadriremes and embarked himself . . . He hurried to a place from which others were fleeing, and held his course directly into danger. Was he afraid? It seems not, as he kept up a continuous observation of the various movements and shapes of that evil cloud, dictating what he saw.

Ash was falling onto the ships now, darker and denser the closer they went. Now it was bits of pumice, and rocks that were blackened and burned and shattered by the fire. Now the sea is shoal; debris from the mountain blocks the shore. He paused for a moment wondering whether to turn back as the helmsman urged him. "Fortune helps the brave," he said. "Head for Pomponianus."

At Stabiae, on the other side of the bay formed by the gradually curving shore, Pomponianus had loaded up his ships even before the danger arrived, though it was visible and indeed extremely close, once it intensified. He planned to put out as soon as the contrary wind let up. That very wind carried my uncle right in, and he embraced the frightened man and gave him comfort and courage. In order to lessen the other's fear by showing his own unconcern he asked to be taken to the baths. He bathed and dined, carefree or at least appearing so (which is equally impressive). Meanwhile, broad sheets of flame were lighting up many parts of Vesuvius; their light and brightness were the more vivid for the darkness of the night. To alleviate people's fears my uncle claimed that the flames came from the deserted homes of farmers who had left in a panic with the hearth fires still alight. Then he rested, and gave every indication of actually sleeping; people who passed by his door heard his snores, which were rather resonant since he was a heavy man. The ground outside his room rose so high with the mixture of ash and stones that if he had spent any more time there escape would have been impossible. He got up and came out, restoring himself to Pomponianus and the others who had been unable to sleep. They discussed what to do, whether to remain under cover or to try the open air. The buildings were being rocked by a series of strong tremors, and appeared to have come loose from their foundations and to be sliding this way and that. Outside, however, there was danger from the rocks that were coming down, light and fire-consumed as these bits of pumice were. Weighing the relative dangers they chose the outdoors; in my uncle's case it was a rational decision, others just chose the alternative that frightened them the least.

They tied pillows on top of their heads as protection against the shower of rock. It was daylight now elsewhere in the world, but there the darkness was darker and thicker than any night. But they had torches and other lights. They decided to go down to the shore, to see from close up if anything was possible by sea. But it remained as rough and uncooperative as before. Resting in the shade of a sail he drank once or twice from the cold water he had asked for. Then came a smell of sulfur, announcing the flames, and the flames themselves, sending others into flight but reviving him. Supported by two small slaves he stood up, and immediately collapsed. As I understand it, his breathing was obstructed by the dust-laden air, and his innards, which were never strong and often blocked or upset, simply shut down. When daylight came again 2 days after he died, his body was found untouched, unharmed, in the clothing that he had had on. He looked more asleep than dead. (Pliny the Younger 1963)

In a second letter to Tacitus a few days later, Pliny wrote:

By now it was dawn, but the light was still dim and faint. The buildings round us were already tottering, and the open space we were in was too small for us not to be in real and imminent danger if the house collapsed. This finally decided us to leave the town. We were followed by a panic-stricken mob of people wanting to act on someone else's decision in preference to their own (a point in which fear looks like prudence), who hurried us on our way by pressing hard behind in a dense crowd. Once beyond the buildings we stopped, and there we had some extraordinary experiences which thoroughly alarmed us. The carriages we had ordered to be brought out began to run in different directions though the ground was quite level, and would not remain stationary even when wedged with stones. We also saw the sea sucked away and apparently forced back by the earthquake: at any rate it receded from the shore so that quantities of sea creatures were left stranded on dry sand. On the landward side a fearful black cloud was rent by forked and quivering bursts of flame, and parted to reveal great tongues of fire, like flashes of lightning magnified in size . . .

Soon afterwards the cloud sank down to earth and covered the sea; it had already blotted out Capri and hidden the promontory of Misenum from sight. Then my mother implored, entreated and commanded me to escape the best I could—a young man might escape, whereas she was old and slow and could die in peace as long as she had not been the cause of my death too. I refused to save myself without her, and grasping her hand forced her to quicken her pace. She gave in reluctantly, blaming herself for delaying me. Ashes were already falling, not as yet very thickly. I looked round: a dense black cloud was coming up behind us, spreading over the earth like a flood. "Let us leave the road while we can still see," I said, "or we shall be knocked down and trampled underfoot in the dark by the crowd behind." We had scarcely sat down to rest when darkness fell, not the dark of a moonless or cloudy night, but as if the lamp had been put out in a closed room. You could hear the shrieks of women, the wailing of infants, and the shouting of men; some were calling their parents, others their children or their wives, trying to recognize them by their voices. People bewailed their own fate or that of their relatives, and there were some who prayed for death in their terror of dying. Many besought the aid of the gods, but still more imagined there were no gods left, and that the universe was plunged into eternal darkness for evermore. There were people, too, who added to the real perils by inventing fictitious dangers: some reported that part of Misenum had collapsed or another part was on fire, and though their tales were false they found others to believe them. A gleam of light returned, but we took this to be a warning of the approaching flames rather than daylight. However, the flames remained some distance off; then darkness came on once more and ashes began to fall again, this time in heavy showers. We rose from time to time and shook them off, otherwise we should have been buried and crushed beneath their weight. I could boast that not a groan or cry of fear escaped me in these perils, had I not derived some poor consolation in my mortal lot from the belief that the whole world was dying with me and I with it.

At last the darkness thinned and dispersed into smoke or cloud; then there was genuine daylight, and the sun actually shone out, but yellowish as it is during an eclipse. We were terrified to see everything changed, buried deep in ashes like snowdrifts. We returned to Misenum where we attended to our physical needs as best we could, and then spent an anxious night alternating between hope and fear. Fear predominated, for the earthquakes went on, and several hysterical individuals made their own and other people's calamities seem ludicrous in comparison with their frightful predictions. But even then, in spite of the dangers we had been through, and were still expecting, my mother and I had still no intention of leaving until we had news of my uncle. (Pliny the Younger 1963)

Pompeii was buried under more than 20 m (66 feet) of ash. It was abandoned and long forgotten as new towns were later built above the buried ruins. Then in 1748, well diggers accidentally rediscovered the ruins of Pompeii (plate 5A). For more than two centuries, it has been gradually and nearly completely excavated, to reveal a picture of everyday life in ancient Rome, complete with frescoes and graffiti on the walls and the tools and implements of daily life. Among the most remarkable discoveries were cavities in the ash that were encountered during digging. When the cavities were filled with plaster, they formed molds of the bodies of Romans who had died in the ash and vaporized, leaving the hollow behind (plate 5B). Hundreds of these bodies were found, usually in poses suggesting agonizing deaths, asphyxiated in a few seconds by volcanic ash and gases, curling up to shelter themselves. Most of the population of 20,000 apparently died instantly. Only a few thousand survivors reached the safety of boats.

Herculaneum was buried under an even thicker blanket of hard tuff. It was rediscovered in 1709. Excavation began in 1738, but digging has been difficult and it is only partly exposed. Unlike Pompeii, Herculaneum was a smaller town (about 5,000) but a rich coastal resort, with a more affluent population, as shown by the clothes and jewelry that have been found. Archaeologists found not only the cavities left by bodies but also 300 skeletons in death poses. They were found near the waterfront attempting to escape and were apparently killed by superheated volcanic gases before they too were vaporized, leaving only bones.

After the great eruption of AD 79, Vesuvius returned to a more active phase, with frequent eruptions nearly every century for another two millennia. The eruption of AD 203 was recorded by the historian Cassius Dio, and in AD 472, its ash reached as far as Constantinople. The most destructive recent eruption was in 1906 when a record number of lava flows appeared and 100 people died. The most recent eruption in 1944 (fig. 3.1) destroyed many villages as well as 88 World War II B-25 bombers that were fighting farther north on the Italian peninsula. For the past 67 years, Vesuvius has been relatively dormant, although plenty of steam still rises from the crater, and there are many small earthquakes. But given its history, it is still considered one of the world's most active and dangerous volcanoes. Despite these problems more than 3 million people live all around its base, and 1 million live on the actual slope of the volcano.

Krakatau, West of Java

Although the eruption of Vesuvius is the most famous volcanic disaster in European history, its eruptions are small compared with some of the huge explosive events elsewhere. Perhaps the most famous of these is the eruption of Krakatau (the Indonesian spelling, although it has been anglicized to "Krakatoa") volcano in the Sunda Straits west of Java and south of Sumatra. (In 1969, Hollywood made a dud of a disaster movie about the volcano, full of scientific inaccuracies, not the least of which was the erroneous title, *Krakatoa, East of Java*). In 1883, the Indonesian archipelago was part of the hugely profitable

Krakatau, Indonesia

Fig. 3.2. Changes in the shape of Krakatau before and after the eruption. The three major peaks of the original volcano (Rakata, Danan, and Perboewatan) are shown by triangles; two of them vanished when the mountain exploded. (Map courtesy USGS; redrawn by Pat Linse)

colony of the Dutch East Indies. Dutch planters and colonialists occupied many large cities and controlled a huge native population, producing nearly all the world's supply of pepper and quinine, a third of its rubber, a quarter of its coconuts, and a fifth of its tea, cocoa, sugar, coffee, oil, and other commodities. Heavy ship traffic passed through the Sunda Straits to Batavia (now Djakarta) on the north shore of Java and between the Indian and Pacific oceans.

Although the volcanoes of Krakatau had been dormant for more than 200 years, by 1882 and 1883, numerous earthquakes had occurred in the area. On May 20, 1883, ship captains reported seeing plumes of steam rising from Perboewatan, the northernmost of the island's three cones (fig. 3.2). Soon ash shot out of the vent to an altitude of 6 km (20,000 feet), and the explosions could be heard in Batavia, more than 160 km (100 miles) away. Activity died down for a few weeks, only to resume on June 16 with more explosions and a thick ash cloud that darkened the skies over Indonesia for five days. A wind arose on

June 24 revealing two ash plumes, both from vents near the center of the volcano and south of Perboewatan. This eruption increased the tidal range, forcing crews to chain ships to the docks. Many earthquakes followed, and passing
ships reported large rafts of pumice floating in the Indian Ocean to the west.

On August 11, H. J. G. Ferzenaar landed on the island and conducted the
only study of the early phase of activity before the final cataclysm. All vegetation had been burned down to tree stumps, and there were now three major
ash plumes, with the new one coming from the central peak of Danan. Steam
plumes rose from 11 other vents, and there was an ash layer 0.5 m (20 inches)
deep across the entire island. Ferzenaar reported these discoveries and recom-
mended that no one else visit the island, although life still continued as normal in distant cities like Batavia.

The vents continued to spew ash, pumice, and steam for many more days.
The next big eruption occurred on August 25, which spewed a cloud of black
ash 27 km (17 miles) high, with explosions occurring about every 10 minutes.
Hot ash and pumice up to 10 cm (3 inches) in diameter covered the decks of
ships 20 km (11 miles) away. One of the volcanic cones collapsed into the sea,
triggering a small tsunami that hit the shores of Java and Sumatra up to 40 km
(28 miles) away. The eruption was so loud that people in far off Batavia could
not sleep, as the deafening pounding noise kept up hour after hour.

The final paroxysm occurred on August 27, when four large explosions occurred between 5:30 a.m. and 10:41 a.m. (plate 6). They were so loud that they
were heard in Perth, Australia, more than 3,500 km (2,200 miles) away, and
even on the island of Mauritius north of Madagascar, 4,800 km (3,000 miles)
away, where they sounded like nearby cannon fire. These explosions generated a pressure wave that ruptured the eardrums of sailors in the Sunda Strait.
The barometric pressure gauges measuring normal air pressure jumped off the
scale and were shattered. This aerial shock wave continued around the globe
and registered on barographs worldwide up to seven days after the initial explosions. Burning ashes rained down over the region, killing about 1,000 people
in the village of Ketimbang more than 40 km (25 miles) north of Krakatau, and
incinerating nearly all the forests around the Sunda Strait. Research in the

past few decades has shown that pyroclastic flows can move across the water's surface because of their intense heat and speed, buoyed by a frictionless layer of gases beneath them (like the puck in an air hockey game). This explains why they could spread across the ocean from the volcano to more distant islands and shores. More than 3,000 people on the island of Sebesi were killed with no survivors, even though they were 13 km (8 miles) from Krakatau. The most deadly consequence, however, was the gigantic tsunamis and the collapse of the caldera, which killed at least 36,000 and maybe as many as 126,000 (see chapter 2).

The wife of town controller Janni Beyerinck described her experiences in the village of Ketimbang, where the pyroclastic flows trapped many and killed most of the residents; however, a few were spared:

> Suddenly, it became pitch dark. The last thing I saw was the ash being pushed up through the cracks in the floorboards, like a fountain. I turned to my husband and heard him say in despair "Where is the knife? . . . I will cut all our wrists and then we shall be released from our suffering sooner." The knife could not be found. I felt a heavy pressure, throwing me to the ground. Then it seemed as if all the air was being sucked away and I could not breathe . . . I felt people rolling over me . . . No sound came from my husband or children . . . I remember thinking, I want to . . . go outside . . . but I could not straighten my back . . . I tottered, doubled up, to the door . . . I forced myself through the opening . . . I tripped and fell. I realized the ash was hot and I tried to protect my face with my hands. The hot bite of the pumice pricked like needles . . . Without thinking, I walked hopefully forward. Had I been in my right mind, I would have understood what a dangerous thing it was to . . . plunge into the hellish darkness . . . I ran up against . . . branches and did not even think of avoiding them. I entangled myself more and more . . . My hair got caught up . . . I noticed for the first time that [my] skin was hanging off everywhere, thick and moist from the ash stuck to it. Thinking it must be dirty, I wanted to pull bits of skin off, but that was still more painful . . . I did not know I had been burnt. (Scarth 1999)

By the next morning on August 28, Krakatau was eerily quiet. Small mud eruptions occurred, but major volcanic activity had ended. Before the volcano, Krakatau had three major cones. Afterward, a huge, drowned caldera 250 m (850 feet) deep with only three small islands remained of the outer base of the volcano (fig. 3.2). About 18–21 km^3 (4.3–5.0 cubic miles) of volcanic material (mostly ignimbrites formed by hot flowing ash and gases) had been displaced, deposited over an area of 1.1 million km^2 (420,000 square miles). Most of this material filled the basins of the Sunda Strait around Krakatau or covered the shorelines of Java and Sumatra and the adjacent islands. There were rafts of pumice floating on the Indian Ocean, some with human skeletons on top. A few of these rafts washed all the way to the east coast of Africa. The huge amount of ash shot into the stratosphere blocked sunlight and dropped global average temperatures about 1°–2°C for more than a year. Weather patterns were erratic for years, and temperatures did not return to normal until 1888. The skies dimmed, even darkened for months after the eruption, and high amounts of particulate matter in the stratosphere changed the color of the sky, producing spectacular orange red sunsets as depicted in Edvard Munch's famous 1893 painting *The Scream*. Munch described the sky over Norway after the eruption as follows: "Suddenly the sky turned blood red . . . I stood there shaking with fear and felt an endless scream passing through nature" (Prideaux 2005, 17). Rare atmospheric effects, such as a blue moon, a Bishop's ring around the sun in daytime, and volcanic purple light at twilight were also seen around the world.

Krakatau remained quiet for a few more decades until submarine eruptions were detected in the center of the caldera on December 27, 1927. A few days later, this eruption built a new, much smaller volcano on top of the old vent known as Anak Krakatau ("son of Krakatau"). Most of the initial ash and pumice produced by this eruption were quickly eroded away by the sea until 1930, when a series of lava flows built up the island faster than the sea could erode it. Anak Krakatau has continued to grow about 13 cm (5 inches) per week since the current phase of eruptions began in the 1950s. It goes through phases of activity in which it builds rapidly, then several years of quiescence, followed by renewed eruptions. The most recent eruption was in April 2008, releasing hot gases, rocks, and lava in all directions.

Although the Dutch had ruled the area for centuries, by 1883 the Dutch East Indies colony was in an advanced state of decay, with continual uprisings by the indigenous peoples forcing the Netherlands to suppress revolts in one war after another. Simon Winchester (2003) argued that the eruption of Krakatau helped weaken the Dutch East Indian colony even further by crippling its economy and exposing the incompetence and corruption of the Dutch administrators. It triggered a wave of anti-Western militancy in the native Muslim populations. According to Winchester, these events sped up the eventual demise of the Dutch colonies, which finally ended when the Japanese invaded in January 1942. When Japan surrendered in 1945, the Dutch were unable to regain control of the islands, and by 1949 Indonesia was independent.

Although the Krakatau eruption was spectacular, it wasn't even the biggest eruption in Indonesia in the past 200 years. (Indonesia has hundreds of active volcanoes, the most of any country in the world.) That distinction goes to the eruption of Mount Tambora, on the island of Sumbawa, east of Java. When it erupted in 1815, it was the largest eruption in recorded history. The explosion was heard on Sumatra, more than 2,000 km (1,200 miles) away, and heavy ash falls covered most of Indonesia. It killed between 70,000 and 90,000 people, 12,000 of whom died from the eruption; the rest starved because crops were destroyed.

After the eruption, the most definitive description of the volcano and its aftermath was given by Sir Thomas Stamford Raffles, a governor-general of the Dutch East Indies, the founder of Singapore, and a respected naturalist. He wrote:

Island of Sumbawa, 1815

In April, 1815, one of the most frightful eruptions recorded in history occurred in the mountain Tambora, in the island of Sumbawa. It began on the 5th day of April, and was most violent on the 11th and 12th, and did not entirely cease till July. The sound of the explosion was heard in Sumatra, at a distance of nine hundred and seventy geographical miles in a direct line, and at Ternate, in an opposite direction, at the distance of seven hundred and twenty miles. Out of a population of twelve thousand, only twenty-six individuals survived on the island. Violent whirlwinds carried up men, horses, cattle, and whatever else came within their influence, into the air, tore up the largest trees by the roots, and covered the whole sea with floating timber. Great tracts of land were covered by lava, several streams of which, issuing from the crater of the Tambora mountain, reached the sea. So heavy was the fall of ashes, that they broke into the Resident's house in Bima, forty miles east of the volcano, and rendered it, as well as many other dwellings in the town, uninhabitable. On the side of Java, the ashes were carried to the distance of three hundred miles, and two hundred and seventeen towards Celebes, in sufficient quantity to darken the air. The floating cinders to the westward of Sumatra formed, on the 12th of April, a mass two feet thick and several miles in extent, through which ships with difficulty forced their way. The darkness occasioned in the daytime by the ashes in Java was so profound, that nothing equal to it was ever witnessed in the darkest night. Although this volcanic dust, when it fell, was an impalpable powder, it was of considerable weight; when compressed, a pint of it weighing twelve ounces and three quarters. Along the sea-coast of Sumbawa, and the adjacent isles, the sea rose suddenly to the height of from two to twelve feet, a great wave rushing up the estuaries, and then suddenly subsiding. Although the wind at Bima was still during the whole time, the sea rolled in upon the shore, and filled the lower parts of houses with water a foot deep. Every prow and boat was forced from the anchorage and driven on shore. The area over which tremulous noises and other volcanic effects extended was one thousand English miles in circumference, including the whole of the Molucca Islands, Java, a considerable portion of Celebes, Sumatra and Borneo. In the island of Amboyna, in the same month and year, the ground opened, threw out water, and closed again. (Raffles 1830)

Tambora had worldwide effects as well. It injected so much dust into the stratosphere that the earth's weather patterns changed, and dust blocked sunlight from reaching the surface. The year 1816 came to be known as the "Year without Summer" because it produced cold, dark rainy summer months in North America and in Eurasia. This led to crop failures, livestock starvation, and widespread disease (especially a typhus epidemic) and famine among populations around the world. In June, it snowed in New York, in New England, and in many European cities. In Lord Byron's villa near Lake Geneva in Switzerland, Percy and Mary Shelley were visiting and told gothic horror stories to pass the time in the cold, dark, wet summer. That gloomy time inspired Mary Shelley to write *Frankenstein* (1818).

Winds of Fire: Mont Pelée, 1902

One of the first scientifically documented instances of *nuées ardentes*, or pyroclastic flows, occurred during the 1902 eruption of Mont Pelée on the Caribbean island of Martinique. At that time, Martinique was a busy, heavily populated island, with a large port city, St. Pierre, home to more than 25,000 people. Mont Pelée had been dormant for centuries when on April 25, 1902, it came to life with geyser-like clouds of ash and steam known as fumaroles, followed a day later by showers of ash and cinders. Each day was marked by more ash falls and other evidence that an eruption was about to happen. About 700 rural folks fled to safety in the city each day. But there was an election scheduled for May 10, and the governor did not want people fleeing the region and changing the results. He commanded the militia to preserve order and turn back people fleeing the city, a fatal decision for him and inhabitants of St. Pierre. On May 2, the mountain produced loud explosions, earthquakes, and a massive pillar of ash, which covered the landscape for miles and darkened the sky for hours. Then the wind changed, the ash was blown northward, and it was thought that the worst was over.

On May 5, a host of bizarre events foretold doom to come. Distressed and agitated livestock tried to jump their fences. At a sugar mill north of the city, yellowish-speckled ants and foot-long black centipedes, driven from the slopes of Mont Pelée by constant tremors and ash, swarmed over the ground. They bit animals and people, but no one died from these attacks. Snakes invaded the streets of St. Pierre, including poisonous pit vipers, 2 m (6 feet) long, known as the *fer de lance*. They bit pigs, chickens, and people who did not get out of their way, killing more than 200 animals and 50 people. The mayor of St. Pierre ordered soldiers to shoot them, and at least 100 snakes were killed.

On May 8, after several more days of active eruption, the mountain suddenly exploded (fig. 3.3A), and a dense superheated ash cloud rushed down the mountainside at 670 km/h (420 mph), engulfing and burying St. Pierre in less than a minute. Anything flammable was incinerated by the 1,000°F heat. More than 30,000 people died instantly. Sailors who dove into the water as their ships blazed were badly burned and did not survive. A warship approaching from the sea was forced to wait offshore for several days because of the intense heat of the pyroclastic material. When the fires finally burned out, only rubble and ruin were left of this big city (fig. 3.3B). Everything else had been destroyed. Only one man survived in the city. Prisoner Louis-Auguste Cyparis had been confined in a deep dungeon. He was rescued four days later, and for years after, he made a living as the "Lone Survivor of St. Pierre," a sideshow attraction in the Barnum and Bailey circus. The only other survivor was a young shoemaker, Léon Compere-Léandre, who lived on the outskirts of the city and escaped the worst of the flows. His description of the ordeal follows:

I felt a terrible wind blowing, the earth began to tremble, and the sky suddenly became dark. I turned to go into the house, with great difficulty climbed the three or four steps that separated me from my room, and felt my arms and legs burning, also my body. I dropped upon a table. At this moment four others sought refuge in my room, crying and writhing with pain, although their garments showed no sign of having been touched by flame. At the end of 10 minutes one of these, the young Delavaud girl, aged about 10 years, fell dead; the others left. I got up and went to another room, where I found the father Delavaud, still clothed and lying on the bed, dead. He was purple and inflated, but the clothing was intact. Crazed and almost overcome, I threw myself on a bed, inert and awaiting death. My senses returned to me in perhaps an hour, when I beheld the roof burning. With sufficient strength left, my legs bleeding and

Fig. 3.3. A, Eruption of Mt. Pelée in 1902, showing the *nuées ardentes* flowing down to the sea, and ash plumes ejected to the stratosphere. B, Ruins of St. Pierre. (Photos by Angelo Heilprin, 1902. Courtesy Wikimedia Commons.)

covered with burns, I ran to Fonds-Sait-Denis, six kilometers from Saint-Pierre. (Pellegrino 1999, 299)

Why Are There Different Kinds of Eruptions?

Not all volcanoes explode like Vesuvius, Krakatau, Tambora, or Mont Pelée. Hawaiian volcanoes, familiar from nature films, erupt with relatively thin lava flows but do not blow their tops in a catastrophic explosion. Why the difference? The answer lies in the chemistry of the magma. Volcanoes like those on Hawaii or Iceland erupt *basaltic* lavas, which are composed of material similar to the composition of mantle from which their magma comes. It is rich in magnesium, iron, and calcium, and relatively depleted in silica, aluminum, potassium, and sodium. Basalts are made of olivine (also known as the gemstone peridot), pyroxenes, and calcium-rich plagioclase feldspar. These minerals have very high melting temperatures (typically 1,600°C), so they form fluid, hot magmas. Basaltic lava flows like water. When these volcanoes erupt, they release seemingly endless lava flows. These volcanoes do not explode, except for throwing some small lava blobs into the air from lava fountains.

The lavas from the explosive eruptions of Krakatau or Mont Pelée have different magma chemistry. These lavas are known as *andesites, dacites,* and *rhyolites* and are richer in silica, aluminum, potassium, and sodium but were depleted in chemicals found in basalts, such as iron, magnesium, and calcium. This chemistry forms a completely different suite of minerals, especially quartz (silicon dioxide), the black flaky mica biotite, amphiboles like hornblende, and both potassium-rich feldspar and sodium-rich plagioclase feldspar. These magmas form at much lower temperatures (as low as 600°C) than do basalts.

Why are andesite-dacite-rhyolite magmas chemically transformed from their original basaltic mantle sources? The answer is complicated, requiring a detailed understanding of mineralogy and geochemistry. Simply put, magmas change chemistry as they melt their way to the surface from their original source. They can be changed by a number of processes, including contamination by melting silica-rich crustal rocks from the walls of the magma chamber, melt-

ing of previously cooled magma chambers that release low-temperature minerals such as potassium feldspar and quartz first (*partial melting*), and depletion of high-temperature minerals that settled out of the magma chamber early
and change the chemistry of the remaining melt (*fractional crystallization*).

Consequently, andesite-dacite-rhyolite magmas are not only different in
chemistry and mineralogy but also in behavior. They are viscous, like sticky
peanut butter or molasses. Rather than flow freely like basalts, andesite-rhyolite
lavas flow slowly or not at all. If the throat of one of these volcanoes is choked
by a sticky, viscous dacite or rhyolite plug, the mountain will build up pressure until it explodes. This produces the full range of pyroclastic materials:
ash, pumice, hot, glowing clouds of volcanic material (*nuées ardentes*, or pyroclastic flows), thick tuff deposits, and so on.

Andesite-dacite-rhyolite magmas are found primarily on the Pacific Ring of
Fire (see fig. 1.6), the Mediterranean, and a few other places. As established in
chapter 1, these places are subduction zones, where one tectonic plate sinks
beneath another (fig. 3.4). As one slab goes down into the mantle, the sinking
slab begins to melt, releasing some of its material into magmas that then must
melt their way up through the overlying crust (already composed of silica-rich
materials) until they erupt to form subduction volcanoes (known as "arc volcanoes" because many are part of island arcs like the Aleutians or Indonesia).
Most volcanoes in the world formed in a subduction setting (especially in
the Ring of Fire from the Andes to the Cascades to the Aleutians to Japan, the
Philippines, and Indonesia) are generated this way.

By contrast, volcanoes that erupt basaltic lavas, with their deep mantle
sources, come from different plate tectonic settings (fig. 3.4). Some, like Hawaii, Yellowstone, and Iceland, lie above "hot spots" in the earth's mantle,
where molten plumes of mantle material constantly melt their way to the
surface. Others, like the midocean ridges and Iceland, lie above places where
the crust is pulling apart in a spreading zone, and fresh basaltic magma wells
up to fill the cracks. Most of these ridge volcanoes are small submarine eruptions, although some rise high enough to form islands (or in the case of Iceland, a hot spot lies directly below the midocean ridge).

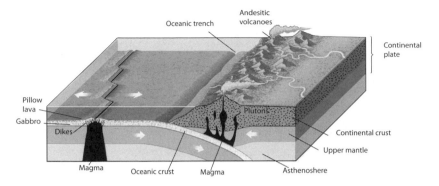

Fig. 3.4. Plate tectonic settings of volcanic activity. (Image courtesy USGS, redrawn by Pat Linse)

Still others are formed when huge cracks open up in the continental crust, which allow mantle-derived magma to flow up to the surface. These tend to produce long fissures and rifts from which pour immense sheets of lava that cover the landscape in days. Not surprisingly, these are known as "flood basalts." Such eruptions happened 15–16 million years ago when the Columbia River basalts erupted and flowed across all of eastern Oregon and Washington and parts of Nevada and Idaho. They covered some 130,000 km² (50,000 square miles), with hundreds of flows totaling more than 1,500 m (5,000 feet) in total thickness, with individual flows up to 100 m (330 feet) thick. These eruptions produced more than 104,000 km³ (25,000 cubic miles) of lava. These lava flows, or the thin soils that developed atop them, can be seen everywhere in eastern Washington and Oregon. Similar gigantic basalt floods occurred in Siberia 250 million years ago, in western India and Pakistan 65 million years ago, and in many submarine eruptions that produced features like the Ontong-Java Plateau and the Kergeulen Plateau, which are still under the ocean.

Basaltic volcanoes are not nearly as deadly as andesite-dacite-rhyolite volcanoes. Although eruptions on Hawaii or Iceland can produce lava flows that will incinerate everything in their paths, they move slowly, enabling people to escape, unless they behave foolishly or are trapped. Certainly, the landscapes of Washington and Oregon before the flood basalt eruptions were devastated.

In the the famous Gingko Petrified Forest near Vantage, Washington, trees that grew on soils atop old lava flows were fossilized when another lava flow covered the forest and entombed them. In another instance near Grand Coulee, Washington, a bloated rhinoceros carcass floating upside down in a lake was surrounded by lava flows that molded its shape as lava cooled. Life was abundant on the Columbia River landscapes 15 million years ago, but it was incinerated each time a new eruption occurred.

Many buildings and forests in Hawaii and Iceland have been wiped out by basaltic lava flows. These eruptions did not explode and produce ash that covered huge areas in hours, circled the globe, or cooled the planet, as andesite-dacite-rhyolite volcanoes such as Krakatau, Tambora, or Vesuvius.

Given these simple but critical distinctions, we can now see that basaltic volcanoes are different from andesite-dacite-rhyolite in their magma chemistry and mineralogy, magma behavior, temperature, viscosity, and location on the earth's surface. Giant fast-moving lava flows come from basaltic eruptions, whereas explosive pyroclastics, ash, and pumice come largely from andesite-dacite-rhyolite eruptions, and they are usually not found together. Silly Hollywood movies, such as *Volcano* and *Dante's Peak*, are all that more ridiculous because they throw every possible volcanic event together in a great illogical mixture that would never actually happen in nature (not that Hollywood movies have *ever* been scientifically accurate).

Supervolcanoes

One geologic term that does come from television is *supervolcano*, coined by a BBC program in 2000. The Discovery Channel remade the film in 2005 with several other documentary channels. Geologists have still not formally accepted this neologism, but it describes many eruptions in the prehistoric past beyond the scale of anything witnessed in recorded history (although some were within the time that prehistoric humans lived). By definition, supervolcanoes erupt material in excess of 1,000 km^3 of ejecta and reach VEI 8 on the scale of volcanoes ("VEI" stands for volcanic explosivity index).

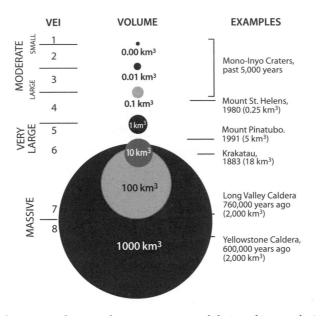

Fig. 3.5. Relative size of some volcanic eruptions, and their ranking on the VEI (volcanic explosivity index) scale. (Courtesy USGS, redrawn by Pat Linse)

Only a handful of eruptions have so far been identified as supervolcanoes (fig. 3.5). Even though Mount St. Helens received tremendous publicity when it erupted in 1980 (mostly because it was in a populated area of the United States and not in the less-developed world), it was a dwarf compared with most other great eruptions. Even Krakatau and Tambora are beneath the threshold for supervolcanoes. About 75,000 years ago, an eruption occurred on Mount Toba in Sumatra, which ejected 2,800 km³ of material. It was believed to be the largest volcanic explosion in the last 25 million years. It released the energy equivalent of 1 gigaton of explosives, forty times larger than the largest nuclear bomb explosion, and about 3,000 times as powerful as the eruption of Mount St. Helens. Toba injected so much ash into the stratosphere that the ash clouds blocked the sun's radiation and decreased global temperature by 3°–5°C (5°–9°F), further amplifying the cold of the ongoing ice ages. The tree line and snow line dropped 3,000 m (9,000 feet) lower than today, making

most high elevations uninhabitable. Global mean temperatures dropped to
only 15°C after three years and took a full decade to recover to pre-eruption
temperatures. Ice cores from Greenland show evidence of this dramatic cooling in the trapped ash and ancient air bubbles, although so far it has not been
detected in Antarctic ice cores.

A number of geneticists and archaeologists have argued that the Toba catastrophe nearly wiped out the human race, leaving a genetic bottleneck of only
about 1,000 to 10,000 breeding pairs of humans worldwide. In a genetic bottleneck, there are so few breeding pairs that they do not have the full complement
of genes of their ancestors, and their descendants have a different genetic code
as well. In addition to the geologic evidence of Toba's size and atmospheric
effects, geneticists have found evidence from the molecular clocks in our ge-
nomes that human populations were in a genetic bottleneck at about this
time. One other study found a similar genetic bottleneck in the genes of human
lice, and in our gut bacterium *Helicobacter pylori*, which causes human ul-
cers; both of these date back to the time of Toba, according to their molecular
clocks. The Toba catastrophe theory is still debated, but it is reasonable to
think that such a global catastrophe would have profound effects on the
human population.

Although Indonesia has a lion's share of the world's deadliest volcanoes,
North America was home to many supervolcano eruptions. The caldera in the
heart of Yellowstone National Park, this nation's first national park, is still a
threat. The hot springs, geysers, and fumaroles of Yellowstone indicate that
an enormous mantle hot spot lies not far beneath the crust, and Yellowstone's
ring shape is actually due to the collapse of a giant volcano into a caldera.
Based on the dates of the volcanic flows all around Yellowstone, it last erupted
about 640,000 years ago, ejecting more than 1,000 km³ of material that blanketed the entire western half of the United States. This deposit is known as the
Lava Creek tuff. Thin layers of this ash can be detected in ice age deposits
as far east as Louisiana and Alabama, more than 3,000 km (1,000 miles) away
(fig. 3.6). Another large eruption from the same hot spot occurred on the Idaho
side of Yellowstone, when the Island Park caldera exploded about 2.1 million

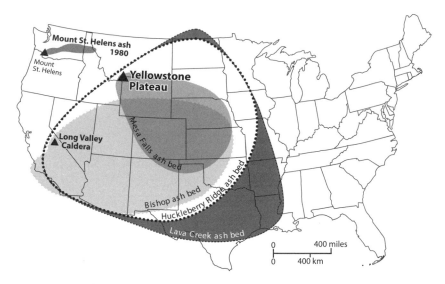

Fig. 3.6. Map of the distribution of volcanic ash from some of the larger prehistoric eruptions. (Courtesy USGS, redrawn by Pat Linse)

years ago and spewed more than 2,500 km³ of ash and pumice known as the Huckleberry Ridge tuff (fig. 3.6), which blew east as far as Missouri and as far south as Texas.

Another active but slightly smaller eruptive source is the Long Valley caldera, near Bishop and Mammoth Mountain, on the east flank of the Sierras in California (fig. 3.6). About 730,000 years ago, it exploded and shot 600 km³ of material into the air, which blanketed the entire western half of the United States as far east as Oklahoma. Its biggest eruption is known as the Bishop tuff, which is found all around the northern Owens Valley and can be traced to ice age outcrops in Kansas and Nebraska.

These eruptions are dwarfed by the eruption of the La Garita caldera in southwestern Colorado. It is a huge hole in the ground, 35 km by 75 km (22 miles by 47 miles), which forms a large irregular oblong valley in the San Juan Mountains. Some of the caldera and its ash deposits can be seen in the Wheeler Geologic Area northwest of South Fork, Colorado (fig. 3.7). In other places, the caldera has been covered up and partially filled by eruptions of the

Fig. 3.7. Many layered tuff deposits, eroded into pinnacles, from the Wheeler Geologic area. (Photo courtesy G. LeVelle)

smaller but younger Creede caldera and other younger calderas. When La Garita exploded 27.8 million years ago, it blew more than 5,000 km^3 (1,200 cubic miles) of material all over the United States, enough material to fill Lake Michigan. It deposited a widespread ash layer known as the Fish Canyon tuff, which can still be seen in the Arkansas River Canyon, 100 km northeast of the caldera, and beneath the surface of the Alamosa area, 100 km east of the volcano. At one time, it was one of the largest volcanic deposits in the world and probably covered much of the Rockies and the western and central plains of the United States. It is calculated that the energy released by the La Garita explosion was 10,000 times more powerful than the largest nuclear device that humans have ever detonated.

La Garita appears to have been the largest explosive volcanic eruption documented so far, but it pales in comparison to the eruption of the Siberian basalt flows. These occurred about 250 million years ago, in concert with the biggest mass extinction of all time. Although these events did not produce ash clouds that covered the planet, they did release huge amounts of mantle gases that may have triggered a massive "super-greenhouse" global warming event. The thickness and volume of lava flows is truly staggering, estimated to be almost 4 million cubic kilometers of lavas, or about a thousand times as large as any of the eruptions discussed so far. The implications of this eruption will be addressed in chapter 12.

Volcano Warning—and Protection?

As the case histories have shown, most volcanoes explode with little or no warning, and when they explode, people in the immediate surrounding region have a small chance of survival. As soon as volcanoes become active and show signs of fumaroles and small earthquakes, it is time to evacuate areas closest to the mountain. Unfortunately, many of the historic eruptions before the birth of modern volcanology (such as Vesuvius, Krakatoa, and Mont Pelée) were not taken seriously, and many people died who should have evacuated. In many parts of the world, governments employ volcanologists to monitor volcanic activity and to give them warnings and recommendations. Some countries (especially Indonesia and the United States) that have frequent eruptions have researched their volcanoes and usually can influence the proper authorities to close and evacuate areas. In countries with little investment in geological research, however, government authorities may not understand the seriousness of the volcanic threat or may resent foreign intervention and ignore volcanologists' recommendations, resulting in unnecessary loss of life when the eruption does come. Chapter 4 looks at the classic example of how Colombia failed to heed the warnings of volcanologists in the 1985 eruption of Nevado del Ruiz, resulting in thousands of deaths.

Volcanologists' predictions are not easy or always precise. Many volcanoes show signs of life, accompanied by earthquakes, fumaroles, and even small explosive eruptions of ash, and then go dormant. When volcanologists order expensive evacuations and then no catastrophe strikes, ungrateful displaced people are resentful and may sue. As with earthquake prediction, predicting volcanoes can never be perfect because of our limited understanding of volcanoes and because not all volcanoes behave alike.

Still, there have been many successes. Most of the big eruptions in Indonesia in the past few decades were much less deadly because the government has taken volcanology seriously, and people heed warnings to evacuate. In 1980, Mount St. Helens erupted after plenty of warnings, so that most people were evacuated from a large area surrounding the volcano. The only exceptions were geologists who thought they were a safe distance away, and a few hard-headed individuals such as Harry Truman, a crusty old owner of Spirit Lake Lodge. Truman loved the media attention he received for thumbing his nose at authorities and declaring that he would die rather than leave his beloved forest. Unfortunately, he got his wish. Fifty-seven people lost their lives because Mount St. Helens surprised geologists by exploding in a giant sideways blast toward the north, rather than straight up. People closer to the volcano on the south, west, and east sides were not killed, but people up to 37 km (23 miles) to the north, farther than anyone expected, were killed. A well-known victim was Dave Johnston, a 30-year-old U.S. Geological Survey geologist, who had been monitoring the volcano for months. Johnston was the only geologist to predict correctly that Mount St. Helens would erupt with a lateral rather than vertical blast, based on studies of deposits from its past eruptions. His research pioneered the use of volcanic gases to predict an imminent eruption. Even he thought that his observation post 10 km (6 miles) to the north was at a safe distance when on the morning of May 18, 1980, the mountain's north flank blew outward. His last recorded words were his excited radio message to his base in Vancouver, Washington, "Vancouver! Vancouver! This is it!" His body was never found, although the charred remains of his trailer were discovered in 1993.

Probably the most successful prediction of a serious volcanic event was the eruption of Mount Pinatubo in the Philippines in 1991, the second largest volcanic eruption of the twentieth century. It is one of hundreds of volcanoes in that island nation, due west of Manila on the island of Luzon, near Subic Bay Naval Base and Clark Air Force Base, American military bases that date back to the U.S. colonial era. Mount Pinatubo had been dormant in historic times (see *Nova* documentary, "In the Path of a Killer Volcano"), but on July 16, 1990, a 7.7 magnitude earthquake struck to the north, devastating the entire region, which may have triggered the upwelling of magma into Pinatubo. On March 15, 1991, Pinatubo awoke, with fumaroles of steam, fissures opening on the summit, and many small earthquakes. Geologists had mapped and dated the ancient deposits around the mountain and realized that it was likely to erupt in a huge explosion, with ash sheets and volcanic mudflows (*lahars*, which are discussed in chapter 4) on a gigantic scale. The U.S. Geological Survey monitored the activity from Clark Air Force Base and used Air Force planes and helicopters to observe the ash plumes, to take samples on the summit, to measure gases emitted by the vent, and to follow the eruption stages from seismographs placed all around the summit.

On June 3, the first significant magmatic eruption occurred, followed by a large explosion on June 7. The Philippine government heeded the U.S. and Filipino geologists and evacuated the area up to 20 km away from the summit. The mountain went through phases of quiescence and renewed activity, causing stress among geologists who rode the roller coaster of daily change and worried that premature warnings might lead to cynicism and discredit them. By June 14, geologists detected serious signs that the volcano was ready to blow and issued an alert to everyone within 40 km of the volcano, displacing 60,000 residents. Finally, tiltmeters showed that the volcano was becoming inflated, so geologists warned nearly everyone to evacuate, and most of the multi-million-dollar aircraft left Clark Air Force Base. Shortly after noon on June 14, seismographs suddenly stopped sending signals as they were incinerated by the first pyroclastic flows, and the mountain blew its top, turning the day into night and showering the nearby area with a rain of pumice and ash (made worse by a typhoon that struck at the same time). The last brave geologists and military personnel escaped as quickly as they could, and all survived their close call with death. Even though millions of people once lived near the volcano, only about 800 died, mostly killed by houses collapsing under the weight of wet volcanic ash, not by the blast. Economic misery for the Philip-

Fig. 3.8. The eruption of Eldfell volcano above the town buried houses to the eaves in ashes and cinders. (Photo courtesy USGS)

pines went on for years though thousands survived. Both Subic Bay and Clark bases were too damaged for the United States to fix. With the end of the Cold War and Philippine resentment of their presence, the United States closed the bases and pulled out of the country permanently.

All of these andesite-dacite-rhyolite volcanoes are deadly and unpredictable, although geologists are getting better at it. Because of the size of volcanoes and explosive capacity, evacuating—not preventing their effects—is the only option. The less explosive basaltic volcanoes are different. Hawaii regularly experiences lava flows, yet almost no one dies, because lava is easier to outrun; major losses occur to fixed features, such as buildings, roads, and plantations. In one case, humans actually stopped volcanic eruptions from wreaking devastation.

On January 23, 1973, volcanic fissures opened up outside the fishing town of Heimaey, an island off the south coast of Iceland. Lava flows and a huge plume of cinders poured into the hastily evacuated town, but all 5,300 residents escaped, although the town was devastated (fig. 3.8). As the volcano (now named Eldfell) grew larger, it began to issue lava flows that spread around its base. It was soon clear that the flows might block off the harbor and destroy any chance that the town would be able to survive without its harbor and fishing fleet. So the Icelanders took action. They brought in a large number of hoses and pumps and blasted seawater at the front of the advancing lava flows, chilling them down and stopping them. This took several months throughout February and March, but eventually the eruption slowed and the lava flow stopped. The flows partly blocked the mouth of the harbor, creating a natural breakwater that improved the harbor. Normally, humans are at the mercy of volcanoes, but once in a while, it is possible for humans to stop nature actually in its tracks.

FOR FURTHER READING

Decker, R., and B. Decker. 1981. *Volcanoes.* W. H. Freeman, New York.

McPhee, J. 1989. Cooling the lava. In *The Control of Nature.* Farrar, Straus and Giroux, New York.

Scarth, A. 1999. *Vulcan's Fury: Man against the Volcano.* Yale University Press, New Haven, CT.

Walker, B. 1982. *Volcanoes.* Time-Life Books, Chicago.

Winchester, S. 2003. *Krakatoa: The Day the World Exploded, 27 August 1883.* HarperCollins, New York.

Zeilinga de Boer, J., and D. T. Sanders. 2002. *Volcanoes in Human History: The Far-Reaching Effects of Major Eruptions.* Princeton University Press, Princeton, NJ.

Landslides

Gravity Always Wins

The mountain is coming down!" he shrieked. Standing directly underneath
the tumbling hillside, hearing its terrible crackle and roar and watching
a plume of earth spew toward the sky, the men broke and ran for their lives . . .
He lost sight of the others as the hillside bore down. Out of the corner
of one eye, he could see a house and a trailer in hot pursuit.
—*Los Angeles Times*, January 2005

A Surfer's Nightmare: La Conchita, California, 2005

The winter of 2004–2005 was wet in Southern California. Many places in the
steep mountains behind the urban belt had flooded and experienced landslides. Huge amounts of rain had fallen in the last weeks of December and the
first weeks of January. In the sleepy coastal town of La Conchita, there was no
reason to think that the winter rainy season would be unlike any other. La
Conchita consisted of a few dozen houses with about 300 residents, located
right on the coast on Highway 101 between the wealthier communities of Santa
Barbara and Ventura. La Conchita was much more laid-back and inexpensive,

Fig. 4.1. The scar and earthflow deposits of the 2005 La Conchita landslide. The slide covered most of the town at the base of the slope. (Photo by the author)

with small beach cottages inhabited mostly by retired surfers, artists, beach-combers, and hippies who savored their pleasant beachfront life without Santa Barbara's high prices and congestion.

It was also located north of the freeway and railroad tracks, at the base of a steep cliff made of loose sandstones and mudstones once deposited in ancient seas (fig. 4.1). These ancient sedimentary rocks were then uplifted by faults to heights more than 150 m (500 feet) above sea level. The cliffs were prone to landslides up and down the coast, from Malibu to the sea cliffs west of Ventura. They were formed of softer sedimentary rocks, with lots of clays that soaked up water, expanded, and became slippery when they were saturated. The bluffs above the town had many landslide scars, showing a long history of instability. In March 1995, part of the hillside had given way, covering up the houses on the street against the base of the cliff. The 1995 slide measured 120 m (400 feet) wide, 330 m (1,100 feet) long, and spread across 4 hectares

(10 acres). It was greater than 30 m (100 feet) deep, with an estimated volume 1.3 million cubic meters (1.7 million cubic yards). This had also been a wet year, with 390 mm (15 inches) of rain in the two weeks before the slide. No one was killed in the event, because the landslide moved relatively slowly and residents were warned to evacuate, but a number of houses were destroyed. Afterward, large retaining walls were built at the base of the slide debris to stop further movement.

Still, the residents of La Conchita were pretty mellow during the rains of 2004–2005 and did not expect anything different from previous rainy winters. Early on the morning of January 10, small mudflows started moving down the nearby canyons. Highway 101 was closed, and emergency officials and TV crews monitored the mudflows. Then at 12:30 in the afternoon, the cliff gave way, as the TV news cameras were rolling. (Video footage of this event is online and amazing to watch.) A mass of earth 350 m (1,150 feet) long and 80–100 m (260–330 feet) wide, remobilized from the unstable 1995 landslide, quickly moved downslope at 10 m/sec (33 feet/sec), faster than anyone could outrun it. Before people in the houses below could react, it had overrun almost half the town, burying and destroying 13 houses, damaging 23 others, and burying dozens of people. Many of the houses were pushed forward as if by a mighty bulldozer and then torn apart before being buried. Emergency workers and townspeople rushed to aid people trapped under their smashed houses and rescued quite a few. Still, some houses were so deeply buried and crushed that there was no way to dig down without rescuers endangering their own lives, and there were no sounds or other signs of life. The rescuers eventually gave up, and these bodies remain buried in the slide mass, with memorials marking the site where they vanished (plate 7).

Many witness accounts were recorded, including this story from the *Los Angeles Times*:

> Standing directly underneath the tumbling hillside, hearing its terrible crackle and roar and watching a plume of earth spew toward the sky, the men broke and ran for their lives . . . Instinctively, he ducked for cover, throwing himself

between two cars and wedging into the smallest of lifesaving cracks, just as the mud and debris washed over him. "I just turned and dove underneath these cars, that's what saved my life," said Ray from the county hospital in Ventura, with scratches and bruises and a gaping wound where a splintered 2-by-4 had speared his right leg. (*Los Angeles Times*, 2005)

After the landslide, residents rebuilt their lives and mourned their dead. Many blamed other people for the landslide, not Mother Nature and not their own decision to live in this community. Some residents filed lawsuits against the landowners on the top of the cliff, blaming the landslide on irrigation of the croplands above the cliff. Their suits were thrown out of court because geologic evidence clearly indicated that the extraordinary rains and the cliff's unstable nature caused the landslide. Many people urged the remaining residents to sell and move elsewhere, so that the town could be bulldozed and made into a park. Because of the steep drop in housing values, no one wanted to sell at a loss. Besides, La Conchita was one of the last relatively inexpensive communities on that stretch of coast. Today, town residents are still in denial, unable or unwilling to move out. They leave memorials (plate 7) along the fence at the edge of the slide mass. Neither the state nor anyone else has the money or willpower to buy out residents and relocate them out of harm's way, so they stay, waiting for the next landslide to wipe out even more of the town and claim more lives.

Black Monster: Debris Flows of the San Gabriel Mountains

The San Gabriel Mountains rise to the north of Los Angeles to summits more than 3,100 m (10,000 feet) above the valleys below, one of the steepest mountain fronts in the world. They are also one of the fastest-rising ranges in the world, with average rates of uplift of several millimeters per year, although they tend to move much more than during the frequent earthquakes, when the mountain front may jump upward and southward a meter or more. Their steep relief and high elevation are due to active faults along the range, from the San Andreas fault on the northern edge, to the San Gabriel fault, down the middle to the Sierra Madre–Cucamonga thrust fault that plunges beneath the range from the southern foothills.

Their steepness combined with heavily shattered rocks make them prone to landslides. In addition, most of Southern California is near-desert that receives about 25 cm (10 inches) of rainfall in a normal year and only a few inches in drought years. It is bone dry nearly year round, with only a handful of large, wet Pacific storms to provide rains in the winter months. When these storms arrive, they slam up against the mountains and drop amazing amounts of rainfall in a short time. In the winter storms of 2004–2005, more than 1.42 m (56 inches) of rain fell in Pasadena, California. One of the most intense rainstorms ever recorded comes from my own backyard, the Rossmoyne station in north Glendale (Daingerfield 1938). In a storm that started February 27 and ended March 4, 1938, more than 330 mm (3 feet) of rain fell in one week, with a maximum intensity over one 5-minute period of 122 mm/h (roughly 5 inches an hour). This is still one of the most intense rainstorms ever recorded in the world. These intense cloudbursts quickly overwhelm the natural array of channels and canyons, and pour down as huge floods choked with mud and boulders known as debris flow. It has the viscosity of wet concrete, and it is so dense and fast moving that it can carry huge boulders, as well as lighter objects like fire engines, coffins, cars, and houses. Debris flows are particularly common in wet winter months right after a brushfire has burned the slopes. Not only are the slopes free of vegetation to hold back water and soil but the burned chaparral brush produces creosote oils that permeate the soil and waterproof it; water flows off the surface rather than sinking in.

Debris flows happen frequently in the canyons along the San Gabriel front. Pulitzer Prize–winning author John McPhee's terrifying example is recounted in the essay "L.A. against the Mountains":

In Los Angeles versus the San Gabriel Mountains, it is not always clear which side is losing. For example, the Genofiles, Bob and Jackie, can claim to have lost and won . . . On a February night some years ago [1978], the Genofiles were awakened by a crash of thunder and lightning striking the mountain front.

Ordinarily, in their quiet neighborhood, only the creek beside them was likely to make much sound, dropping steeply out of Shields Canyon on its way to the Los Angeles River. The creek, like every component of all the river systems across the city from mountains to ocean, had not been left to nature. Its banks were concrete. Its bed was concrete. When boulders were running there, they sounded like a rolling freight train. On a night like this, the boulders should have been running. The creek should have been a torrent. Its unnatural sound was unnaturally absent. There was, and had been, a lot of rain . . . The San Gabriels, in their state of tectonic youth, are rising as rapidly as any range on earth . . . Shedding, spalling, self-destructing, they are disintegrating at a rate that is also among the fastest in the world. The phalanxed communities of Los Angeles have pushed themselves hard against these mountains, an aggression that requires a deep defense budget to contend with the results . . . Shields Creek passes under the street, and there a kink in its concrete profile had been plugged by a six-foot boulder. Hence the silence of the creek. The water was now spreading over the street. It descended in heavy sheets. As the young Genofiles and their mother glimpsed it in the all but total darkness, the scene was suddenly illuminated by a blue electrical flash. In the blue light they saw a massive blackness, moving. It was not a landslide, not a mudslide, not a rock avalanche; nor by any means was it the front of a conventional flood. In Jackie's words, "It was just one big black thing coming at us, rolling, with a lot of water in front of it, pushing the water, this big black thing. It was just one big black hill coming toward us."

In geology, it would be known as a debris flow. Debris flows amass in stream valleys, and more or less resemble fresh concrete. They consist of water mixed with a good deal of solid material, most of which is above sand size . . . The dark material coming toward the Genofiles was not only full of boulders; it was so full of automobiles it was like bread dough mixed with raisins. On its way down Pine Cone Road, it plucked up cars from driveways and the street. When it crashed into the Genofiles' house, the shattering of safety glass made terrific explosive sounds. A door burst open. Mud and boulders poured into the hall. . . . The house became buried to the eaves. Boulders sat on the roof [fig. 4.2A].

Fig. 4.2. A, Photograph of a house in La Crescenta, California, buried to the rooftop in February 1978. Workmen stand on the roof and inspect a car that was buried on top of the house. The chimney for the stove can be seen on the right. (Courtesy USGS) B, A house destroyed during the February 5–6, 2010, debris flows at the top of Ocean View Boulevard in La Cañada, California. The mud and debris flowed over a wall of concrete K-rails and tore through the front walls of the house, completely filling it and destroying everything inside. Here, the salvage crews have already removed the crushed cars and major debris piles from the front of the house and are now clearing out the interior before it is demolished. (Photo by the author)

Thirteen automobiles were packed around the building, including five in the pool. A din of rocks kept banging against them . . . The house had filled up in six minutes and the mud stopped rising near the children's chins. (McPhee 1989, 184–86)

The events of the 1978 debris flows were nearly repeated in the same neighborhood 30 years later. In late August and early September 2009, the Station Fire burned almost 200,000 acres of the Angeles National Forest, including all the hillsides above the neighborhood where my family and I live. We choked back ash and endured the 105°F heat wave as we moved boxes into our new home. Even though there were no Santa Ana winds to push the fire, it burned for weeks, fueled by many decades' worth of accumulated dry brush, and could not be stopped. Hundreds of hard-working fire crews managed to save nearly all the homes against the mountain front. Then came warnings that the barren slopes would be sliding downhill when the rains came.

Sure enough, in mid-January we had four days of heavy rains (more than 12 inches in the foothills) that saturated the hillsides and caused some sliding; however, most of the debris was contained in the huge debris basins in the canyons, designed to catch rocks and mud but allow the water to flow downhill. When the rain stopped, dump trucks traveled up and down the steep mountainside streets to empty debris basins before the next rains.

On the night of February 5 and 6, 2010, a rainstorm (predicted to be relatively small) reached our foothill area. It became an intense thunderstorm, with howling winds and pounding rain. My rain gauge measured 6 inches of rain in a few hours. The already saturated hillsides soon started to move, and huge debris flows came roaring down Ocean View Boulevard in La Cañada, tumbling cars and turning them into twisted wrecks and lifting 20-foot 8,000-pound pieces of concrete barriers known as K-rails, tossing and hurling them into other objects. These huge barricades had been put up since the Station Fire to divert and slow the river of mud, but they were no match for the power of debris flows. Most residents had evacuated at the first sign of rain, but the few who stayed behind despite the warnings were hunkered down in their homes because escape through the roiling river of mud, rocks, and cars was impossible. Emergency services chose not to send them a reverse-911 evacuation call (where the police department 911 computers send a call to everyone in a phone grid) because they were safer indoors than if they tried to evacuate through the dangerous storm.

Nevertheless, danger to the homes was real. Even though it had been cleared since the January rains, the 23-foot-deep Mullally Canyon debris basin overflowed in a few hours during the night. The houses at the top of Ocean View Boulevard were overwhelmed when the monstrous black walls of mud and rocks burst through the K-rails, fences, and sandbags and through windows, doors, and walls, filling homes with mud that reached the ceiling. In some places, the bent and twisted cars acted like battering rams, smashing through walls. Some homes were destroyed (fig. 4.2B), and at least 15 were red-tagged by authorities and residents were not allowed inside for fear they could collapse.

As reported by Thomas Curwen in the *Los Angeles Times* (February 12, 2010):

Henry Laguna had never heard a sound so terrifying. Like a train, screeching and crashing as it flew off the tracks. "Get up!" he screamed. He had just left the bedroom to check on the puppy but frantically raced back to wake up his wife and son.

Outside in the darkness, they slipped on the wet concrete. They dashed around the pool to the end of the property, as far from the house as possible . . . Within seconds, a torrent hit them, bringing with it rocks and branches . . .

Henry and his family clung to the fence and the cypress, struggling to keep their footing as the speed of the water and mud picked up. The rain beat down upon them. Debris battered the house. Henry heard glass breaking, wood cracking and snapping. [His wife] Damaris saw a dresser and the couch from the television room slide into their swimming pool.

The report describes how they eventually went back to the house to get out of the chilling downpour and find the dog, but it could not be found. They eventually sought refuge at the house of a neighbor across the street. By day-break, they found their house (fig. 4.2B) destroyed, but they wanted to stay and salvage their belongings. However, the house was too dangerous to walk into, and the hillsides could still slip at any moment, so eventually they were convinced to leave and let the crews finish the demolition of their house and their possessions.

Stories of debris flows are typical around the world. They are enormously powerful and capable of lifting and carrying large amounts of material a significant distance. Some have been clocked at 100 km/h (60 mph), although most are much slower. Houses built against mountain fronts with huge rocks and boulders all around the neighborhood usually indicate that the area was built by catastrophic debris flows and is likely to experience debris flows in the future. Yet people love living in cool mountain canyons and foothills, so they build their homes in the path of disaster, unaware of the stories boulders tell. After major events such as the debris flows of 2010, people demanded that the authorities put in bigger barricades to stop the walls of mud. As geologists have explained, if debris flows can move 8,000-pound K-rails, carry cars and fire engines, and rip houses apart, then no physical barrier will stop them. The only real protection from debris flows is to not live in homes in such dangerous areas in the first place. It's like voluntarily standing in front of a loaded cannon and hoping that it won't fire while you're there.

Earth on the Move

The landslide at La Conchita and the debris flows of the San Gabriels are examples of downslope movement of material, or mass wasting. Whenever land surface is elevated (especially in areas of high relief), gravity is constantly working to move material downhill. A grain of sand on a slope is constantly feeling the pull of gravity, so the only force that prevents it from moving is friction that holds it in place. Once a mass exceeds that frictional threshold, it will start to move.

Many different factors determine whether a slope will stay stable or begin to move. The most important is the steepness, or relief, of the slope—steeper slopes are less stable. If you take dry sand in a sandbox or on the beach and pour it out of a bucket, it will form a little conical hill, and no matter how much more sand you pour, the angle of the slope will reach what is called the angle of repose and get no steeper. Any material that makes it steeper than the angle of repose will slide right off. The angle of repose depends on grain size, so it is very gently sloping in finer sand but can be quite steep in a pile of pebbles or boulders.

Another important factor is the fluid content of the material. Dry sand will never form a slope steeper than the angle of repose for its grain size. If you have built sand castles, you know that if you get the sand just moist enough to stick together, you can build sand castles with vertical or even overhanging walls. In slightly dampened sand, there is just enough water trapped between the air spaces in the sand to form small films like the meniscus of water in a test tube. This film of water has enormous surface tension, so it is an effective binding agent. But as any sand castle builder knows, if the sand is too wet, it will flow rather than pack into a vertical wall. If there is too much fluid, there is so much water in the pore space that there are no air pockets. Thus, there is no surface tension, and the water acts as a lubricant and puts pore pressure on the surrounding grains, moving them apart and promoting flow.

You can apply the sand castle analogy to slopes. Normally, soil and rocks are bound together by the forces that compacted them into rock in the first place as well as clay minerals, cements between the grains, and maybe ground-water to enhance surface tension. When these same materials are oversaturated with water, the pore pressure of the fluids push the grains apart and they are ready to flow. In many cases, specific layers act as lubricating surfaces along which the overlying rocks can flow. This is especially true with layers of shale or mudstone, which are typically made of smectite clays known as montmorillonite. These clay minerals expand dramatically when they absorb water into their crystal lattices. When an entire shale layer becomes saturated, it becomes a layer of slippery mud that can lubricate the motion of huge masses of rock. Then all that is needed is some sort of shock (like an earthquake) to destabilize it and trigger the slide. Sometimes a landslide will start when the oversaturation reaches a critical threshold and the material spontaneously fails and breaks apart.

Such movement occurs in a full range of speeds and behaviors, from slow and imperceptible to the human eye to fast, free-falling with the acceleration of gravity. Creep is the slowest type of movement, in which the top layer of soil moves slowly downhill a few millimeters each year. It is typical of steep slopes with lots of loose soil and moves fastest with frequent freezing and thawing, which lifts and moves the soil particles upward and then downhill. Because the surface moves so slowly, we typically cannot see it in real time, but the movement can be detected by how it displaces stationary objects imbedded in the soil. Fence posts and telephone poles will lean downhill, and tree trunks will tilt downhill at their base but curve to grow vertically to compensate for the movement (fig. 4.3A). Even rocks and soil may show signs of flowing downhill (fig. 4.3B). Creep may not be life threatening, but it is found on virtually every slope in the world, and sometimes moves relatively quickly, destabilizing and tearing apart houses and other structures in a matter of a few years. Consequently, it is responsible for more damage than all other types of downslope movement combined.

Slumps are slightly faster movements, in which a coherent block of earth breaks free from a scarp and slips down the curved face (fig. 4.4). An earthflow occurs if the block of earth breaks up and becomes jumbled or incoherent. Earthflows include mudflows and debris flows, which can move as slowly as 1 mm/day or as fast as 10 km/h (8 mph). If the flow is on steep slopes, especially if it traps a lubricating layer of fluids or gases at the base, it can become a slide and can travel up to 160 km/h (100 mph). Finally, the fastest movement is when an overhanging cliff breaks and rocks drop through the air in free fall. In this case, rocks travel without friction at the acceleration of gravity (9.8 m/sec^2) until they reach bottom and abruptly stop.

Gravity Kills

Fast-moving slides are among the most common and deadly downslope movements. One famous event was the 1963 Vaiont disaster in the southern Alps of

Fig. 4.3. A, Creep has caused these
trees near Devil's Postpile National
Monument to tilt downhill (to the right
in this photo), then grow in an upward
curve to compensate. (Photo by the
author) B, The creep of the vertically
tilted layers near the surface can be
seen by their downhill flexure, as exposed in this roadcut. (Photo courtesy
USGS)

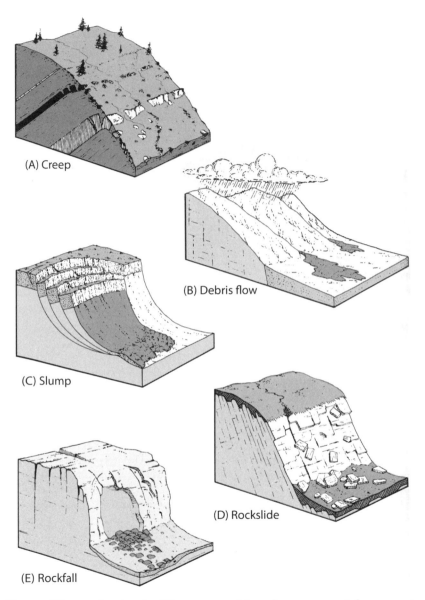

Fig. 4.4. Diagram showing the different types of downslope movement, from creep to rockfalls. (Courtesy USGS, redrawn by Pat Linse)

Italy, about 100 km (60 miles) north of Venice. In 1960, a dam was built across
the Vaiont Valley, impounding 150 million cubic meters (316,000 acre feet) of
water. This water soon saturated in the pore spaces in the shattered rocks from
an old quarry that sat on the flank of the valley. The quarry had layers of rock
that dipped toward the center of the valley, and some of those layers became
slippery when they were water saturated. A number of smaller landslides were
reported, but in response, the Italian government sued the journalists for "un-
dermining the social order" by reporting on this potential danger. Even experiments conducted by the dam's builders showed the kind of disaster that might
occur, but officials refused to accept the results or publicize their discoveries.

In late summer of 1963, heavy rains saturated the shattered rock on the
valley walls. On the night of October 9, 1963, the valley was quiet. Villagers
in Piave River Valley below the dam were inside to avoid the rains and excited
about the evening soccer cup matches. Engineers monitoring the dam began
to worry that a landslide was about to happen and sent a telegram to officials
to warn people; however, no messengers were available to deliver it to its
destination.

At 10:41 p.m., the south side of the Vaiont Valley suddenly broke free, forming a slide mass that was 1.8 km (1.1 miles) long and 1.6 km (1 mile) wide, with
a volume of about 240 million cubic meters. This mass shot down the side of
the valley into the reservoir, filling it with debris up to 150 m (500 feet) above
water level. It continued to shoot up the opposite wall of the valley, and then
back down into the reservoir before stopping (fig. 4.5). It was moving so fast
(110 km/h, or 68 mph) that geologists believed that a carpet of fluids trapped
beneath lubricated it, which made it float freely with minimum friction like a
puck in an air hockey game. This entire process created an earthquake felt all
over Europe and took only 30 seconds. A huge amount of water was displaced
from the reservoir, which sloshed back and forth in what are known as seiche
waves. The dam held up against the pressure, but the sloshing water poured
over the top of the dam at 267 m (780 feet) above valley level, and down into
the Piave Valley. The wall of water rushing down the valley was 70 m (230 feet)
high. It slammed into a number of small towns and killed nearly 3,000 people

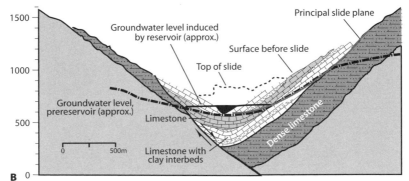

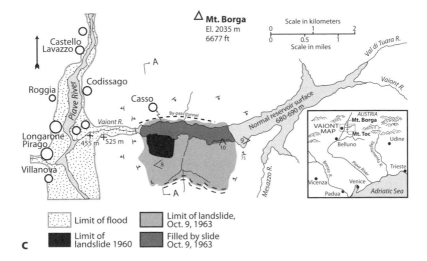

in minutes. It has long been considered the world's worst dam disaster even though the dam held.

At the time it was happening, most people knew nothing of the event because it occurred at night and took out all communications, so no emergency calls for help went out for hours. One survivor, just leaving the bar after the soccer game, felt a blast of cold air as he walked down the street. Then he turned a corner and found everything was gone. The entire town had been washed away or crushed by the boulders carried by the wave of water. He wandered around in the dark all night, trying to find other survivors. In the morning, the valley was covered by a low fog, so there was still no visibility until the fog cleared. Once it did, the town was unrecognizable and was covered in white mud from the limestone-rich sediments washed over by the flood. The survivor then learned that his entire family, including his pregnant wife, was dead.

Another incredible event was the 1925 Gros Ventre slide, just east of Grand Teton National Park in the Gros Ventre Valley of Wyoming. Here the walls are made of heavy blocks of Tensleep Sandstone, which dip steeply into the valley, and sit over dipping clay beds of the Amsden Formation. The rocks are deeply weathered and prone to break up because of heavy freeze-thaw cycles in these high mountains. As early as 1920, parts of the valley wall had begun to slide, with cracks opening up near the top. One local resident, recognizing the danger, sold his farm and got out five years before the disaster. In the summer of 1925, the area was saturated by heavy rains. On the afternoon of July 23, 1925, a huge slide block made of 38.2 million cubic meters (50 million cubic yards) of rock broke free, sliding down more than 640 m (2,100 feet), carrying a dense

Fig. 4.5. (*opposite*) A, Landslide debris looking downstream in the former Vaiont reservoir valley long after the effects of the disaster were over. The small remnants of the lake are visible, but when the disaster occurred, the lake was full, and the landslide debris displaced a huge amount of water. (Courtesy Professor E. Bromhead, University of Kingston) B, Diagram showing the geometry of the Vaiont Valley and the slide. C, Map of the Vaiont Valley, showing the area inundated by the flood and the areas of the 1960 and 1963 slides. (B and C drawn by Pat Linse)

Fig. 4.6. Gros Ventre landslide, Wyoming, June 23, 1925. Across the foot of the new lake to Gros Ventre landslide scar. Lower part of the slide in the background. Slumped east face of the inner side of the dam in the right foreground. (From publications by W. C. Alden. Courtesy USGS.)

pine forest on its top surface intact. Geologists agree that the slide must have trapped a lubricating carpet of air and water beneath it as it moved for it to glide so smoothly over so long a distance without breaking up. When the slide reached bottom, it continued 100 m (330 feet) up the north wall of the valley and then slid back again and formed a debris dam across the river that was 75 m (250 feet) tall (fig. 4.6). This dam completely blocked the river during the rainy fall season, and within three weeks, it had impounded a lake that was 60 m (200 feet) deep with a surface area of 45 km² (11,000 acres).

This was only the beginning of the disaster. The natural rock dam was porous and quickly began to erode, allowing water to seep through. By the spring of 1927, the dam seemed stable in the two years since the slide; most people did not worry about it. However, a heavy snowfall the previous winter was followed by heavy rainfall in May 1927 that accelerated the snowmelt and caused the reservoir level to rise rapidly. By May 18, the rising lake water flowed overtop the dam, releasing 1.2 million liters (43,000 acre feet) of water, producing a 5-m-high (16-foot) wall of water that rushed down the valley and wiped out the town of Kelly only 10 km (6 miles) downstream. Most of the 65 residents survived because they were warned that the dam was about to fail, but 7 people died as they fled with their possessions. Eventually, the escaping water cut a channel through the dam, and the level of Slide Lake is now 15 m (50 feet) lower than the original level, no longer a threat to towns below it.

A Richter magnitude 8 earthquake struck off the coast of Peru on May 31, 1970, to cause one of the deadliest landslides in recent history. The 45 seconds of violent shaking not only destroyed buildings all over the country (with damage reaching Mercalli intensity VIII) but also destabilized a huge mass of rock near the north face of Mount Huascarán in the Andes. This mass of material, a mixture of snow avalanche and debris flow (sometimes called a debris avalanche), sped down the steep mountain valleys at speeds of 160 km/h (100 mph). It was 910 m (3,000 feet) wide and 1.6 km (one mile) long, and picked up additional loose glacial debris from the valley as it moved, eventually producing a sheet of water, mud, and rocks about 61 million cubic meters (2.4 trillion cubic feet) in volume. At that speed, it hit the towns in the valley below within seconds, giving no one a chance to escape. The town of Yungay (fig. 4.7A) was just 18 km (11 miles) downstream and was completely buried in hundreds of meters of debris. Only the steeples and tallest trees protruded from the slide mass. The town of Ranranhirca was also buried to the top. The debris avalanche was so powerful that it picked up buses and trucks and rolled them around like toys (fig. 4.7B) and even carried giant boulders weighing more than 7,000 metric tons (fig. 4.7C). At least 20,000 people in Yungay were buried and died in minutes, and the death toll across Peru exceeded 80,000. Of 400 survivors of Yungay, most happened to be on high ground above the town, including 300 children who were at a circus in a local stadium, which protected them. The earthquake affected an area of 83,000 km² (32,000 square miles), an area larger than Belgium and the Netherlands combined.

An even more amazing event occurred 50 million years ago just east of Yellowstone at Heart Mountain, Wyoming. Heart Mountain (fig. 4.8) is an isolated butte sitting on the rim of the Bighorn Basin. When geologists first studied

Fig. 4.7. The Yungay debris avalanche. A, The village of Yungay, buried under tons of debris. Only the tops of palm trees and the steeples show where the village was located beneath the rubble. B, A tangled mass of buses and trucks tumbled and piled up by the debris avalanche. C, A boulder weighing more than 7,000 metric tons deposited by the debris flow near Ranranhirca. (Photos courtesy USGS)

it, they realized that it was an eroded remnant of a great sheet that slid eastward out of Yellowstone National Park, known as a detachment fault. About 50 to 48 million years ago, a giant sheet of rock about 1,300 km² (500 square miles) in area pulled away from the Beartooth Plateau. It then slid tens of kilometers to the southeast and south into the Bighorn and Absaroka Basins. The gravity slide sheet was composed of a huge mass of limestones overlain by the Eocene Absaroka volcanics. It was originally about 4–5 km thick. Even though the slope was less than 2 degrees, the landslide slid with minimal friction at least 40 km (25 miles); it eventually buried more than 3,400 km² (1,300 square miles). This makes it the largest rockslide ever seen on the surface of the earth.

The great geologic puzzle is how such a huge volume of rock could slide on a nearly horizontal slope with only minimal disturbance and breakage of the slab as it moved. Most geologists believe that the explosion of the Absaroka volcanics must have injected a carpet of hot volcanic gases beneath the slide block, which lubricated its movement so that it encountered no friction (again, much like the puck in an air hockey game, or a hovercraft gliding just above the ground). Models of the movement suggest that it was moving at about 160 km/h (100 mph), traveling roughly 200 miles in 30 minutes.

Killer Lahars

Landslides can be deadly on their own, but when they combine with volcanic activity, they form a terrifying, devastating hot volcanic debris flow. Known by the Indonesian word lahar, they form when a volcanic eruption melts a lot of snow (or water from a summit crater is released) and mixes it with hot volcanic ash and debris. Like a regular debris flow, lahars have the consistency of wet concrete, can carry huge boulders, and can move at speeds approaching 160 km/h (100 mph). Unlike cold debris flows caused by rain, however, lahars are often hot from the volcanic material mixed in. Once they stop, they harden and compact, littered with huge boulders and other debris that they carried.

Nearly every large explosive volcanic eruption that has a source of water produces lahars. Many of the victims of volcanoes discussed in the last chapter,

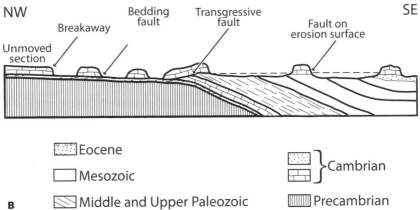

NW   SE

Breakaway — Bedding fault — Transgressive fault — Fault on erosion surface

Unmoved section

Eocene

Mesozoic

Middle and Upper Paleozoic

Cambrian

Precambrian

B

Fig. 4.8. Heart Mountain detachment fault. A, Heart Mountain is an isolated block of Paleozoic limestones sitting atop the glide plane overlying much younger Cretaceous shales (grassy slopes). It is a remnant of a much-larger detachment block that once covered a huge region but is now largely eroded away. (Photo courtesy J. A. Lillegraven) B, Diagram of the Heart Mountain detachment fault, now mostly eroded away except for remnants like Heart Mountain and the McCullough Peaks. (From Pierce 1960, USGS Prof. Paper 400-B: 236–237. Courtesy USGS, redrawn by Pat Linse.)

such as Vesuvius, Tambora, and Krakatau, were buried in lahars that swept across the river valleys. In more recent events, like the 1991 Pinatubo eruption and the 1980 Mount St. Helens eruption, geologists have outstanding film footage of these moving, boiling masses of water and rocks. Geologists have conducted an enormous amount of research into how lahars occur and move. Some ancient lahars are staggering in scale. Geologists studying the valleys below Mount Rainier found evidence of many prehistoric lahars, serious geologic threats to Tacoma and other towns in their path. The Osceola lahar was formed during an eruption of Mt. Rainier 5,600 years ago. It created a wall of mud 140 m (460 feet) deep in White River Canyon and covered an area of more than 330 km² (130 miles). Its total volume is estimated at 2.3 km³ (0.55 cubic miles). The towns around Enumclaw and Buckley, Washington, are built entirely on a valley once covered by the Osceola mudflow, and cities such as Tacoma, Auburn, Sumner, and Puyallup are in the path of future explosive eruptions from Mt. Rainier sending even bigger lahars down the White River.

The deadliest and most devastating recent lahar was formed by the eruption of Nevado del Ruiz volcano in Colombia in 1985 (fig. 4.9). This Andean volcano is more than 5,400 m (17,700 feet) in elevation and was once covered by an ice cap 10–30 m (30–100 feet) thick. In November 1984, the long-dormant volcano awoke and spewed ashes and steam with many small earthquakes. Volcanologists rushed to Colombia to witness the eruption and warned the authorities that the volcano was dangerous, but the volcano kept steadily chugging away without exploding. As the days and weeks wore on, a number of false alarms led to evacuations. Soon the cynical Colombian authorities and residents stopped paying attention and refused to cooperate. As public officials are prone to do, they kept reassuring the people living below the volcano that they were safe.

The mayor of the large city of Armero even assured the citizens of the town that there was nothing to fear. Then at 9:37 that same evening, November 10, 1985, Nevado del Ruiz blew its top with a huge plume of ash and pumice, as well as hot pyroclastic flows that melted nearby glaciers. Mixed with the large volume of melting snow, these events produced multiple lahars that roared

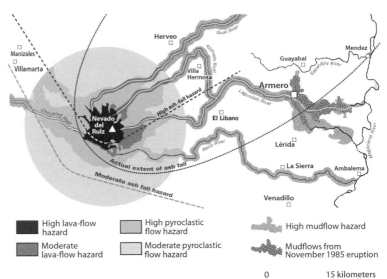

A

High lava-flow hazard
Moderate lava-flow hazard
High pyroclastic flow hazard
Moderate pyroclastic flow hazard
High mudflow hazard
Mudflows from November 1985 eruption

0   15 kilometers

Fig. 4.9. The Nevado del Ruiz eruption and lahars. A, Map of the volcano, and the towns and valleys devastated by the lahar. B, Aerial view of Nevado del Ruiz volcano erupting shortly before the big explosion. C, Aerial view of the city of Armero completely buried in lahar deposits. (Courtesy USGS; A, redrawn by Pat Linse)

Plate 1. A, The Turnagain Heights landslide due to the 1964 earthquake near Anchorage, Alaska, occurred along a steep bluff fronting Knik Arm on Cook Inlet. Its length, which is parallel to the bluff, was about 3 km (1.5 miles); its width was about 0.2–0.4 km (0.25–0.5 miles). Many of the finer homes of the city were reduced to rubble by this landslide. Failure here and in the L Street, Fourth Avenue, and Government Hill landslides in Anchorage occurred on horizontal or near horizontal slip surfaces in the Bootlegger Cove Clay, a marine silt of Pleistocene age. B, Homes damaged by the Turnagain Heights landslide. C, Diagram showing the movement of slump blocks in the Turnagain Heights. (Courtesy USGS; C redrawn by Pat Linse)

Plate 2. A, Uplifted seafloor at Cape Cleare on Montague Island in Prince William Sound in the area of the greatest recorded tectonic uplift on land (10 m, or 33 feet). The very gently sloping, flat rocky surface with the white coating that lies between the cliffs and the water is about a half a kilometer (a quarter of a mile) wide. The white coating consists of the remains of calcareous marine organisms that were killed by desiccation when the wave-cut surface was lifted above high tide during the earthquake. B, Hinchinbrook Coast Guard dock, raised above all but the highest tides by regional uplift in Prince William Sound. Land in this area rose about 3 m (9 feet) during the earthquake. (Photos courtesy USGS Photo Library)

Plate 3. The 1985 Mexico City earthquake showed dramatically how much the vibrating soft lake sediments amplified the shaking and how some buildings with shoddy construction or the right spacing of floors to vibrate sympathetically with the seismic wavelength collapsed, while others next door did not. A, Nuevo Leon fifteen-story reinforced concrete structure. Part of the building was only slightly damaged, while another part of it collapsed. B, Urbana Suarez Apartment Complex completely collapsed. The building next door to it was completely unaffected. (Photos courtesy USGS Photo Library)

Plate 4. A, A shot of the Purtakennedy fishing boat tossed ashore near local businesses in down-town Banda Aceh, Sumatra, Indonesia, following the massive tsunami that struck the area on December 26, 2004. B, Men wandering the beach and helicopters rescuing people offshore, Chennai Beach. (Photo courtesy U.S. Department of Defense)

Plate 5. A, View of Mt. Vesuvius from the ruins of Pompeii. B, Casts of corpses of a group of human victims of the AD 79 eruption of Vesuvius, found in the so-called Garden of the Fugitives in Pompeii. (Photos courtesy Wikimedia Commons)

Plate 6. Ash clouds pouring from the volcano on Krakatau in southwestern Indonesia during the early stages of the eruption, which eventually destroyed most of the island, May 27, 1883. (From the *Royal Society Report on Krakatoa Eruption*, 1888)

Plate 7. The La Conchita landslide of 2005. A, The landslide moved as a rapid debris flow, but it was quite viscous and pushed houses in its path rather than flowing around or through them. The left part of the house was detached from the right part and was pulled by the landslide several decimeters toward the upper left of the photograph. (Photo courtesy USGS) B, Photograph of the memorials placed on the edge of the thick slide mass, commemorating those buried inside. (Photo by the author)

Plate 8. The Big Thompson Canyon Flood, 1976. A, The mouth of Big Thompson Canyon looking upstream into the Narrows. Highway 34 on the left, truncated by the river. B, Cabin lodged on top of a bridge, just below Drake, looking upstream. (Photos courtesy USGS)

Plate 9. Images of Katrina devastation. A, Aerial view of flooding in New Orleans. (Photo courtesy FEMA) B, Image of people stranded on the rooftops by flooding. (Photo courtesy FEMA)

Plate 10. More images of Katrina devastation. A, Downed trees near Slidell, Louisiana. (Photo courtesy NOAA) B, Aerial view of the huge Grand Casino barge in Biloxi, Mississippi, which was originally floating well offshore, but has been blown completely on to the land and across Highway 90 by the hurricane. (Courtesy USGS)

Plate 11. A, Map showing the areas of highest tornado risk. (Diagram courtesy NOAA, redrawn by Pat Linse) B, Damage due to the May 1999 tornado outbreak in central Oklahoma. Aerial shot of the path of the tornado through the suburbs of southwestern Oklahoma City. (Image courtesy NOAA) C, A concrete welcome sign (foreground) lies in front of a devastated Moore, Oklahoma, home. An estimated 1,500 homes in Moore were destroyed. (Image courtesy NOAA)

TORNADO
RISK MAP

Other considerations

≡≡≡ Special Wind Region

▨ Hurricane
Susceptible Region

Wind zones

☐ Zone I
(130 mph)

☐ Zone II
(160 mph)

☐ Zone III
(200 mph)

■ Zone VI
(250 mph)

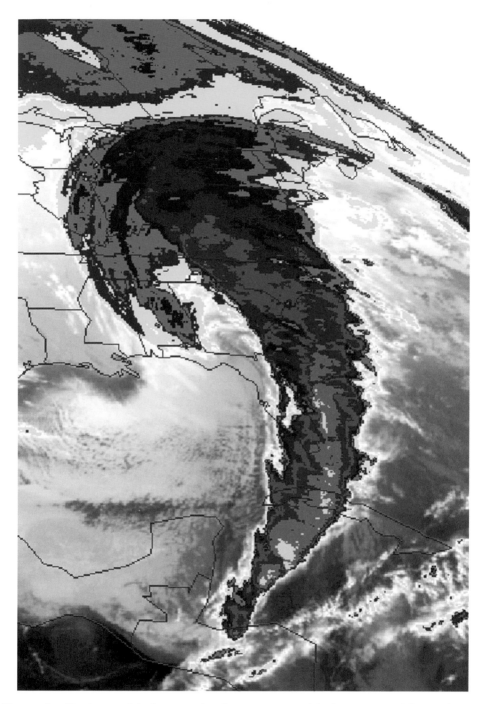

Plate 12. Satellite image of the huge weather front over eastern North America and the Caribbean, March 13, 1993. (Image courtesy of NASA)

Plate 13. Effects of the 1993 "Storm of the Century" in the Deep South. A, Snow in Albertville, Alabama. B, Snow in Christiansburg, Virginia. (Images courtesy National Weather Service)

Plate 14. A. Example of late Precambrian glacial deposits sandwiched between shallow marine tropical limestones. The glacial deposits and cap carbonates in Namibia, with Dan Schrag (*left*) and Paul Hoffman (*right*) for scale. (Photo by G. Walker, courtesy P. Hoffman) B, If the ice sheets return, the Statue of Liberty would just barely be visible above the ice. (Photo courtesy R. Dott)

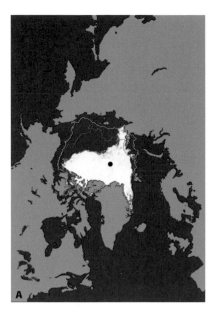

Plate 15. The melting of the Arctic ice cap. A, Diagram showing the expected loss of Arctic ice cap over the next decade. (Images courtesy NASA) B, In the summertime, there is no ice cap over the North Pole, only broken floating ice floes. (Courtesy NOAA Photo Library)

Plate 16. A, Pie chart showing the relative hazard of each natural disaster, expressed as a percentage of the total deaths due to natural disasters. B, Map of the relative risk of death by natural disaster in the United States, organized by counties. The reddest areas are regions of highest risk, the white areas those of intermediate risk, and the bluest areas are those of lowest risk (as measured by standardized mortality ratios). (From Borden and Cutter 2008)

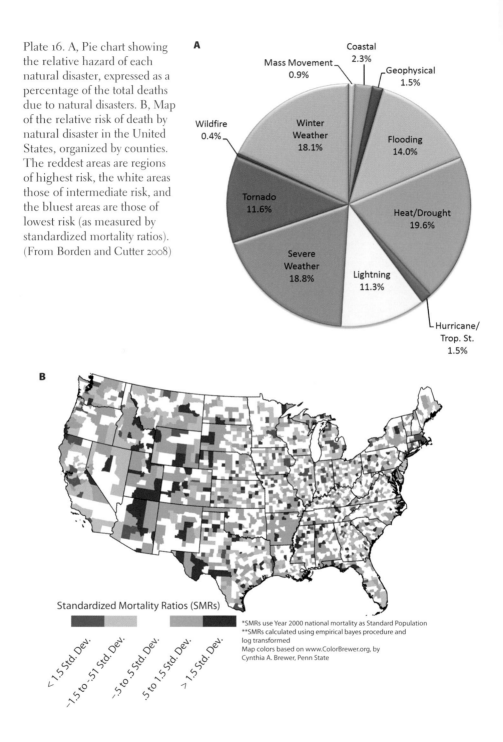

down the valleys around the mountain. First, they wiped out the town of Chin-china, high on the eastern slope, killing 1,800 people. The towns of Libano and Mariquita were also wiped out as the lahars swept through. The lahars flowing down the southern valleys hit a reservoir caused by the landslide dams on the Lagunilla River and water was added to the mix. Some volcanologists compare this wall of water to a megatsunami coming out of the ocean. The wall of water and mud traveled at an estimated 480 km/h (300 mph) down the river valley. By 11 p.m., the lahars had traveled 45 km (28 miles) from their source in the mountains and had dropped more than 5,000 m (16,400 feet) in elevation, giving them a huge gravitational boost. As they reached the broad, flat valleys at the base of the mountain, the lahars slowed to 45 km/h (28 mph), still too fast for anyone to outrun, but the entire town was asleep anyway, re-assured by their mayor and other authorities.

Shortly after 11 p.m., the townspeople awakened to the sound of the roaring lahar. Most, however, could not move quickly enough to outrun it or reach higher ground, and 22,000 residents were buried alive beneath an 8 m (26 foot) thickness of lahar deposits, where they were crushed or suffocated to death. The next morning, rescuers walking across the area could still feel the heat of the lahar. They located bodies by the pools of blood seeping to the surface. A rescue pilot flying over the area in the first daylight remarked, "Oh my God, Armero has been erased from the map" ("Dios mio, Armero ha sido borrado del mapa").

Amazingly, about 5,000 people escaped by fleeing to higher ground. Several people were buried up to their chins in debris but still alive, and rescuers were able to dig them out the next day. Others were not so lucky. Rescuers worked furiously to free people who were buried but found alive. Media from all over the world documented the relief effort. Reporters covered the efforts to rescue a young girl named Omayra Sanchez, who died after three desper-ate days of attempting to free her. After all the work was over and relief had come from all over the world, it was estimated that the disaster cost Colombia $7.7 million, about 20 percent of the country's gross national product for that year. Although it was a natural disaster, the loss of life would have been far less if the authorities had heeded the volcanologists and evacuated people in time.

Sadly, there was additional reason the volcano should have been taken seriously, beyond the volcanologists' warning during the 1985 eruption. The same events had taken place in February 1845, and the same places were buried by the same types of lahars. However, in 1845, the death toll was about 1,000. Few people lived in the mountain valleys of Colombia back then. Geologists had known this history for a long time, but neither past history, nor the evidence from the 1984–1985 precursor to the big eruption, was sufficient to convince the authorities—and 24,000 people died as a result of their negligence.

What Goes Up Must Come Down

Even though the catastrophic deadly events that are the focus of this chapter are dramatic, they are rare. Most of the world's landslides and downslope movement happen gradually and slowly, with little or no loss of life but often with a tremendous loss of property and money. They happen constantly around the world on surfaces that have elevation and relief, and they are unavoidable. The wise will avoid building in places prone to mass wasting. But mountains offer great views and cool, forested canyons, so some will insist on living in the path of danger despite the risk. Sometimes people are unaware of the hazards. Sometimes they simply don't seem to care, or think that with enough money, they can have anything they want. Mountaintop and clifftop homes are often the most expensive and popular because of their views, and their owners seem to think that, as property owners, it is their right that their cliff edge be stable forever and not erode back or collapse. Yet like King Canute ordering the tide not to come in, they are trying to stop natural processes that can only be slowed down temporarily with the infusion of lots of money. Eventually, the slopes and cliffs will come down.

In many cases, people make the situation worse, and then pay a heavy price. It is common practice for developers to carve housing tracts like little shelves jutting out of steep hillsides without regard for the stability of the slope they have cut. They often compound the problem by widening and extending out the housing tract with loose fill dirt from the cut, further destabilizing the slope. Roads and construction often make deep cuts into slopes and mountains to shorten and straighten out their routes. Eventually, nature wins with rockslides and collapsing slopes. We should live with the landscape and build where the ground is stable. Roads should follow natural river valleys (as the pioneers did), or we can continue to try to dominate the landscape to make it fit our convenience. Ultimately, nature will prevail.

The late great political columnist and social commentator Art Buchwald said it best in a column written after one of Southern California's rainy winters and frequent slides:

Los Angeles, a Mobile Society

"Cable," I said, "why do you build your house on the top of a canyon, when you know that during a rainstorm it has a chance of sliding away?"

"It's hard for people who don't live in California to understand how we people out here think. Sure we have floods, and fire and drought, but that's the price you have to pay for living the good life. When Esther and I saw this house, we knew it was a dream come true. It was located right on the tippy top of the hill . . . we could look down on all the smog. Then, after the first mudslide, we found ourselves living next to people. It was an entirely different experience. But by that time we were ready for a change. Now we've slid again and we're in a whole new neighborhood. You can't do that if you live on solid ground." (in Plummer et al., 1999, 209)

FOR FURTHER READING

Brabb, E. E., and B. L. Harrod, eds. 1989. *Landslides: Extent and Economic Significance*. Balkema, Brookfield, VA.

Bruce, V. 2001. *No Apparent Danger: The True Story of the Volcanic Disasters at Galeras and Nevado del Ruiz*. HarperCollins, New York.

Cornforth, D. 2005. *Landslides in Practice: Investigation, Analysis, and Preventative Mediation in Soils*. John Wiley, New York.

Costa, J. E., and G. F. Wierczorek. 1987. *Reviews in Engineering Geology.* Vol. 7, *Debris Flows, Avalanches: Progress, Recognition, and Mitigation.* Geological Society of America, Boulder, CO.

Daingerfield, L. H. 1938. Southern California rain and flood, February 27 to March 4, 1938. *Monthly Weather Review* 1938:139–43.

Dikau, R., D. Brunsden, and L. Schrott, eds. 1996. *Landslide Recognition: Identification, Movement, and Causes.* John Wiley, New York.

Evans, S. G., and J. V. Degraff, eds. 2003. *Catastrophic Landslides: Effects, Occurrence, and Mechanisms.* Geological Society of America, Boulder, CO.

Glade, T., and M. J. Crozier. 2005. *Landslide Hazard and Risk.* John Wiley, New York.

Highland, L. M. 2009. The Landslide Handbook—a Guide to Understanding Landslides. U.S. Geological Survey Circular 1325. http://pubs.usgs.gov/circ/1325/.

Holzer, T. 2009. *Living with Unstable Ground.* American Geological Institute, Alexandria, VA.

Jibson, R. W. 2005. Landslide Hazards at La Conchita, California. U.S. Geological Survey Open File Report 2005-1067. http://pubs.usgs.gov/of/2005/1067/508of05-1067.html.

McPhee, J. 1989. L.A. against the mountains. In *The Control of Nature.* Farrar, Straus and Giroux, New York.

Sassa, K., H. Fukoka, F. Wang, and G. Wang, eds. 2007. *Progress in Landslide Science.* Springer, New York.

Slosson, J. E., A. G. Keene, and J. A. Johnson, eds. 1993. *Reviews in Engineering Geology.* Vol. 9, *Landslides/Landslide Mitigation.* Geological Society of America, Boulder, CO.

Turner, A. K. 1996. Landslides: Investigation and Mitigation. Special Report of the National Research Council 247. Transportation Research Board, Washington, DC.

U.S. Geological Survey. 2009. Landslides Hazards Program Web site. http://landslides.usgs.gov/.

Zaruba, Q., and V. Mencl. 1969. *Landslides and Their Control.* Elsevier, New York.

Floods

Raging Waters

[And Enlil, ruler of the gods said] the earth bellows like a herd of wild oxen.
The clamor of human beings disturbs my sleep. Therefore, I want Adad
[god of the skies] to cause heavy rains to pour down upon the earth,
both day and night. I want a great flood to come like a thief upon the earth,
steal the food of these people and destroy their lives.
—*The Epic of Gilgamesh*, Sumerian, about 2100 BC

The Great Scablands Floods

In the summer of 1922, J Harlen Bretz, a young geologist from the University of Chicago, was mapping the rocks of eastern Washington State. As a high school teacher in Seattle, he had been interested in the origin of the Channeled Scablands ever since 1910, when he saw a newly published topographic map the Potholes Cataract and realized that it resembled a huge Niagara-sized waterfall with almost no water running across it. He drove and hiked back and forth around the Grand Coulee area, mapping and studying the wide steep-walled canyons with only a trickle of water at the bottom. He saw

Fig. 5.1. Artist's conception of the Spokane flood. (Original painting by S. H. Ominski; used by permission)

the evidence of gigantic waterfalls now abandoned, or huge boulders that no stream could move—nor was there any evidence that a glacier had left them there. He saw streams that dropped off plateaus abruptly, rather than carving down to form a smooth slope to their trunk stream. By 1923, he had an outrageous hypothesis to explain these puzzling features. To Bretz, it was clear that a huge catastrophic flood (fig. 5.1), which he called the Spokane flood (fig. 5.2), had flowed across the region, scouring the deep coulees in the bare Columbia River basalts, stripping away the glacial soils (known as the Palouse) that once covered them, and even rounding off the corners of the joints in the basalt flows (fig. 5.3). As far as Bretz was concerned, no normal river could have done these things.

Bretz's first article on the topic was published in 1923 in the University of Chicago's *Journal of Geology*, but it was either ignored or dismissed by most geologists of the time. There were many reasons geologists disagreed with him. Most had never visited these remote areas of eastern Washington, nor had they seen pictures, so they had no real conception of the scale of the great coulees, the degree of water abrasion on the basaltic basement, or the gigantic abandoned waterfalls that Bretz had studied. (His early geologic articles con-

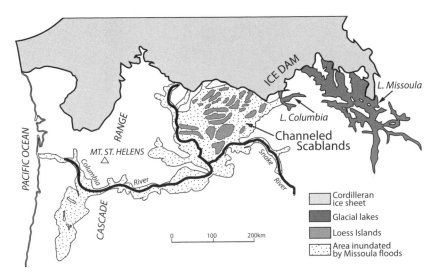

Fig. 5.2. Map of the Pacific Northwest, showing the areas of Channeled Scablands and Glacial Lake Missoula, as well as the ice dam formed by an extension of the Canadian ice sheet. (Courtesy USGS; redrawn by Pat Linse)

tained few photographs, so there was little to help them visualize these features.) In addition, Bretz was a young unknown geologist without the prestigious Ivy League education of the geological elite; therefore, he was not trusted at first. Fundamentally, geologists were skeptical of Bretz's ideas because they smacked of old-fashioned supernatural catastrophes, hearkening back to the bad old days before 1800 when geologists thought that everything was due to Noah's Flood. As discussed in the prologue, pioneering geologists such as James Hutton in the late 1700s, and Charles Lyell in the 1830s had worked hard to expunge the outdated and unscientific practice of ascribing geologic events to untestable, supernatural explanations. Unfortunately, most geologists of Bretz's time followed Lyell's mistake of conflating the uniformity of processes (*actualism*) with the uniformity of rates (*gradualism*) and thus thought that Bretz's floods were unscientific. The big showdown between Bretz and the Establishment took place at the Geological Society meeting in Washington, D.C., on January 27, 1927. Bretz was invited to present his views, only to be ambushed by geologists he called the "six challenging elders," who

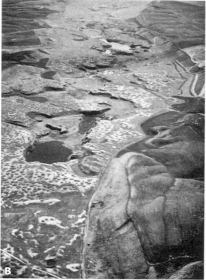

Fig. 5.3. Features of the Channeled Scablands. A, This steep cliff, an abandoned waterfall known as Dry Falls, once had a cataract larger than Niagara Falls, which scoured a deep basin in the basalt beneath it. B, Bedrock channels carved out of basalt (coulees) by the floods (*left*) and uplands that still have their cover of glacial soils (known as Palouse). That the channels on the upper plateaus drop off steeply without a gentle gradient into the main channel was strong evidence that they were cut by water flows that filled the main channel. C, Gigantic ripple marks larger than the houses and barns in the pictures, which are only visible from the air. (Photos courtesy USGS)

offered all sorts of explanations of the puzzling features of the Scablands by conventional river erosion. Bretz gave as good as he got, and stood his ground, but because few people in the audience had actually visited the Scablands, it was impossible to determine the winner of the debate. Another geologist, J. T. Pardee, had worked the area with Bretz and discovered evidence of ancient glacial lake that might explain the Scablands, but he was afraid to speak out on the controversial idea for fear of angering his bosses at the U.S. Geological Survey. One geologist wrote that "the erosional features of the region are so large and bizarre that they defy description." Nevertheless, he remained committed to gradualism. "I believe that the existing features can be explained by assuming normal stream work of the Columbia River . . . Before a theory that requires a seemingly impossible quantity of water is fully accepted, every effort should be made to account for the existing features without employing so violent an assumption" (Allen et al. 1986, 44).

The Scablands flood hypothesis remained in limbo as "unproven" for decades, although Bretz cantankerously defended his idea at every opportunity. Then, after almost 30 years, a new source of data from aerial photos suddenly vindicated Bretz. Once airplanes began to fly over the area and photos were taken, you could see gigantic ripples that were up to 7 m (22 feet) tall and spaced 130 m (425 feet) apart (fig. 5.3C). On air photos, ripples were clear; however, from the ground a geologist might not see them because they are so large that you walk right over them without noticing, and they are covered by sagebrush. By the 1960s, the weight of geologic opinion shifted in Bretz's favor, and more geologists began to see the Scablands in a new light. Hydraulic engineers began to model the scale of the water flow necessary to account for the features of the Scablands. They estimated the maximum discharge capable of moving giant boulders 11 m (36 feet) in diameter. The flooding was caused by a volume of water of about 2,500 km^3 (600 cubic miles) that lasted up to 11 days. This flood volume was more than five times the volume of Lake Erie. The discharge rate is estimated to have been more than 13.7 m^3/sec (484 million feet/second) and traveled at speeds in excess of 30 m/sec (67 mph).

One sticking point remained, which explained geologists' resistance since

Fig. 5.4. Ice age lake terraces on the slopes above Missoula, Montana, relicts of Glacial Lake Missoula, and its multiple episodes of catastrophic drainage. (Photo courtesy USGS)

the 1920s: where did all that water come from? Bretz wasn't really interested in the problem, only the erosional features of the Scablands, so he casually suggested that maybe a volcanic eruption under the glaciers to the north had released the water. His critics quickly pointed out that this explanation was inadequate. Ironically, the source had been identified since the 1880s, and was eventually pointed out by J. T. Pardee. In the slopes above the mountain valleys around western Montana, are multiple horizontal stair step "beaches" caused by ancient glacial lakes that once combined to form Glacial Lake Missoula (figs. 5.2, 5.4). This water was dammed by an extension of a lobe of a glacier over what is now Lake Coeur d'Alene, Idaho, and each time the glacier retreated and the ice dam broke, it catastrophically drained Lake Missoula across the Scablands. It turns out that there was not a single Spokane flood, as Bretz had thought, but multiple floods.

Bretz lived long enough (he died in 1981 at almost 99 years in age) to see his ideas not only vindicated but also established as the foundation for modern planetary geology. As spacecraft brought back images of the surface of Mars, planetary geologists realized that its surface had been eroded by gigantic floods much like those that scoured the Channeled Scablands, and they used Bretz's work to model their own research. Late in his life, Bretz was hailed by planetary geologists as a pioneer in the field. In 1979, at age 96, Bretz received the Penrose Medal of the Geological Society of America, the highest honor in geology, in belated recognition of his brilliant insights. He commented to his son: "All my enemies are dead, so I have no one to gloat over."

Floods, Mythic and Real

Floods have long been an essential part of the mythology of nearly every culture. Since ancient times, nearly every historic civilization was built on floodplains of major rivers: the Tigris-Euphrates, the Nile, the Indus, and the Yellow River of China. The Sumerians and Babylonians describe a great flood in *The Epic of Gilgamesh* and *Enuma Elish*, and these stories were copied by the Hebrews in almost every detail (except for changing the names of the characters) for the two different conflicting stories of Noah's Flood, which are interleaved in alternating verses in Genesis 6. The Greeks had the legend of Deucalion, who repopulated the world after surviving a flood brought on by divine intervention. Similar flood myths appear in Norse, Celtic, Indian, Aztec, Chinese, Mayan, Assyrian, Hopi, Romanian, African, Japanese, and Egyptian mythology (see Prothero 2007, chap. 2, for further details). In most ancient civilizations, sooner or later their floodplain homes and fields were subjected to large ("500-year") floods when unusually wet weather pounded the drainage basins that surrounded them.

In Mesopotamian cultures, big floods not only wiped out buildings and cities and killed many people but also dissolved historic records written on clay tablets. Consequently, the descendants of flood survivors often had little memory, and no written record, of "before the Flood," and the flood myth became a central part of a culture's mythology. Some people have looked at real geologic events, such as the catastrophic flooding of the Black Sea by the

Mediterranean in 5800 BC, and suggested that it might be the source of the Noah's Flood myth. This idea has met with much criticism, but there are always further attempts to find some real basis for particular floods in mythology. However, ever since geology matured in the early 1800s, scientists have abandoned the idea of Noah and one universal flood. A detailed look at the complexity of the geologic record worldwide shows no evidence of a worldwide flood deposit, despite the bizarre, unscientific ideas of creationist "flood geologists" (see Prothero 2007, chapter 3).

Nevertheless, plenty of catastrophic giant floods in human history are not myths. An extraordinarily heavy period of rain in the tributaries of a drainage basin and excess water spills on to the floodplain equal catastrophic floods. For most of human history, people primarily lived on fertile, well-watered floodplains where crops could be grown. Eventually, catastrophic floods washed crops and villages away, and soon thereafter, villagers reclaimed the floodplain to plant again. Ancient Egyptians relied on the annual flooding of the Nile to provide fresh silt and nutrients to grow their crops, and they prayed to their gods to bring the floods on schedule. In 1964, Egypt built the Aswan High Dam in the upper Nile to prevent flooding from washing away permanent towns and villages. The unintended consequence, however, is that the Nile no longer replenishes the soils of the floodplain. Even worse, the low steady flow of the Nile emerging from the dam is insufficient to wash away the salts that form in the soil by evaporation in this hot, dry climate. Now much of Egypt's farmland is too salty to support crops. The dam may have saved some towns, but it was a costly trade-off.

Likewise, the dams built on the Colorado River have prevented flooding downstream, but they too have had unintended consequences. At one time, the annual floods in the Grand Canyon brought down tons of sand and silt from the side canyons to replenish the sand bars and rebuild the rapids. Since the Glen Canyon Dam was completed in 1964, these floods no longer happen. Instead, water is released through the hydroelectric turbines to match the demand for electricity during the blazing summers in Arizona and their daytime air-conditioning use, a procedure known as "peaking." This regular daily pulse of water release behaves like a tide, forcing boaters to tie up their rafts carefully and pitch their camps high on the sandbars so the "tide" does not wash them away in the night. Unfortunately, this practice also erodes the sandbars and destroys the natural habitat of the canyon, which used to be replenished by flooding each year. In 2005, environmental and boaters' groups pressured agencies responsible for the dam to release water to simulate a natural flood. It seemed to work for a short while, but with the many years of drought and the lack of water all over the Colorado River drainage, it will likely not be repeated.

Floodplains Are for Floods

Backwater rising, Southern peoples can't make no time
I said, backwater rising, Southern peoples can't make no time
And I can't get no hearing from that Memphis girl of mine
Water in Arkansas, people screaming in Tennessee
Oh, people screaming in Tennessee
If I don't leave Memphis, backwater spill all over poor me
People, since its raining, it has been for nights and days
People, since its raining, has been for nights and days
Thousands people stands on the hill, looking down where they used to stay
Children stand there screaming: mama, we ain't got no home
Oh, mama we ain't got no home
Papa says to the children, "Backwater left us all alone"
Backwater rising, come in my windows and door
The backwater rising, come in my windows and door
I leave with a prayer in my heart, backwater won't rise no more.
—Blind Lemon Jefferson, "Rising High Water Blues"

In most parts of the world, people and governments have spent millions of dollars to build structures to control the river and prevent floods. Along nearly every major city on the Mississippi, for example, are huge concrete levees with floodgates and other structures to prevent water from inundating the downtown district. However, in a truly massive flood, these confining walls prevent the water from naturally spreading across its former floodplain. Without any other place to go, this confined water then moves downstream and makes the flooding worse in the first area downstream from the levee walls. Once again, nature shows that it cannot really be controlled, and if we try to do so, we will reap the problems in unintended consequences elsewhere.

Mark Twain said of the Mississippi that we "cannot tame that lawless stream, cannot curb or confine it, cannot say to it 'go here or go there,' and make it obey" (Twain 1883, 172). This was vividly demonstrated in the record flooding of the Mississippi Basin in 1993. It all started in the autumn of 1992 when record rains saturated the soils, followed by a heavy winter snow pack. During the spring, the jet stream shifted southward and forced the warm moist air coming up from the Gulf of Mexico to the area over the northern Midwest. This triggered rain over and over again, producing from 400 percent to 750 percent of the usual rainfall for a typical spring in Iowa or Minnesota. Portions of Iowa received 1.2 m (4 feet) of rain in the spring and summer, which flooded the fields and destroyed crops. The ground eventually became saturated; water rose in the Missouri and Mississippi rivers and spread all over their floodplains (figs. 5.5, 5.6). Eventually, even the larger riverfront cities went under water. Davenport, Iowa, was inundated with 4 m (14 feet) of water, completely drowning the downtown area. In Des Moines, Iowa, 250,000 residents lost their drinking water because of groundwater contamination from sewage and fertilizers released by the flooding. In St. Louis, Missouri, the river crested 14 m (47 feet) above flood stage and was barely held back by high floodwalls. Towns that survived the flooding were virtually ruined because of thick mud deposits on the floors and walls, buried crops, and destroyed businesses. Many of the smaller towns along the Missouri and Mississippi, already struggling to keep their people and businesses, were so badly flooded that they had to be abandoned.

Sometimes people do dangerous or foolish things in floods. Not only are there the usual cases of people who refuse to evacuate, or try to drive across flooded roads and are drowned, but even stranger things have occurred. Dur-

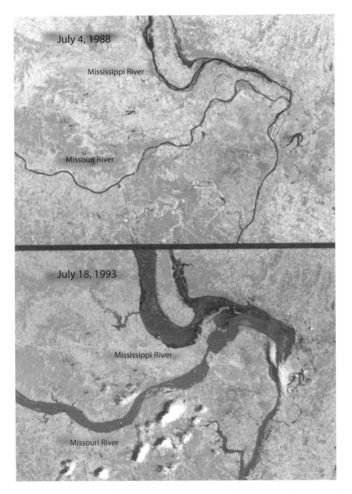

Fig. 5.5. LANDSAT images of the Missouri-Mississippi River confluence before (*top*) and during the peak of the 1993 flooding (*bottom*). Notice the huge expansion of the dark area of drowned floodplains. (Image courtesy NASA)

ing the 1993 floods, 23-year-old James Scott deliberately removed the sandbags from a levee wall to strand his wife on the other side of the river so he could keep on partying. This first breach then led to the failure of more than 1,000 additional levees, and the water quickly got out of control and flooded 57 km² (14,000 acres) of farmland, destroyed many buildings, and closed a bridge. He was sentenced to life imprisonment for his crime.

Fig. 5.6. Aerial shot of the flooding on the Missouri River near Cedar City, Missouri, and north of Jefferson City, Missouri, on July 30, 1993. (Courtesy USGS)

The 1993 flooding continued for 79 days before the waters finally receded. More than 40,000 km² (15,400 square miles) were submerged, 50 people died, and at least 55,000 homes were destroyed, amounting to an estimated $12 billion in damage, although many think it was much higher. It broke the records of the previous monster flood on the Mississippi River of 1927, making it the worst natural disaster in recent U.S. history. In the aftermath, many people called for the government to buy up devastated town sites and to discourage people from rebuilding in the floodplain. After all, *floodplains are for floods*, and we fool ourselves if we build on them thinking that floods will never come. However, most people returned, knowingly putting themselves in harm's way in the hopes that the area would not flood again. In June 2008, heavy rains hit Iowa and the surrounding areas again. Photographs of the drowned cities in 2008 eerily resemble 1993 photos.

Many people who also suffered through the flooding of 1993 had been told by authorities that it was a "100-year-flood" and believed it would not recur for decades, but 2008 proved them wrong. Labels such as "100-year flood" do not indicate that a flood of certain size will happen precisely at 100-year intervals. Instead, it is a statistical average: over an interval of 500 years, or over 100 years, it represents the probability that one flood of a certain size will occur, because there is a strong relationship between the size of a flood and its frequency (fig. 5.7A). As shown in figure 5.7B, the 85-year period from 1908 to 1993 experienced two 100-year-floods with additional floods nearly as large in 1927 and in 2008. These events are unusual, but they are not outside the statistical probabilities calculated for flood occurrences.

Nearly every region of the world has examples of flood stories, such as the great Florence flood of 1966. Florence, Italy, is one of the most beautiful and historic cities in the world. It was largely built during the peak of the Renaissance by the great Italian masters and filled with thousands of amazing works of art produced at that time. I first visited Florence in 1971 and was captivated because I had studied the great masters of art, such as Leonardo da Vinci, Michelangelo, Brunelleschi, Botticelli, Donatello, and such thinkers as Machiavelli, Petrarch, and Boccaccio. I had also written several school reports on the famous Medici family, who had ruled Florence since before the Renaissance and off and on for 300 years afterward.

The city I saw in 1971 was still recovering from severe flooding only five years before. During the fall of 1966, heavy rains had relentlessly pounded the Apennine Mountains, greatly exceeding the normal amount for rainfall at that season. In addition, the region was far more urbanized than it had ever been; therefore, much of the water ran off the concrete and pavement, rather than soaking in. Florence was still built on the old medieval and Renaissance street plan, with little thought to modern drainage and flood control. Some roads actually served as channels for the floodwaters, diverting the water straight to the heart of the city. The narrow Arno River, with its many buildings and bridges, was overbuilt and had no room for water to spread out so that water ended up flooding the city instead. On November 3, 1966, the dams upstream

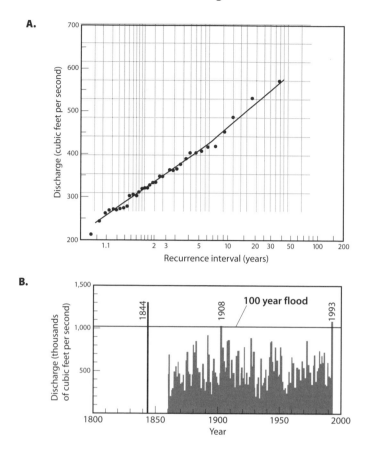

Fig. 5.7. Size-frequency diagram of a typical river system. A, This diagram shows that a flood of certain size is likely to have a recurrence interval of a certain number of years. It does not mean that it takes exactly that same number of years between floods. B, Plot of peak discharge on the Mississippi measured at St. Louis, Missouri. Each bar represents the discharge of a single year, and the horizontal line indicates the threshold for the "100-year-flood." There were two 100-year-floods between 1908 and 1993; these floods were 85 years, not 100 years apart, and there was a third flood almost as large in 2008. (Modified by Pat Linse from USGS)

were overtopped, and released 2,000 m³/sec (70,000 cubic feet per second) toward Florence. Soon the basements began to flood, and a few people started to flee for their lives. At 4:00 the next morning, engineers at Valdano Dam feared it would burst and released as much water as they could down the Arno.

This boiling mass of water reached Florence traveling at 60 km/h (37 mph). Most residents, however, were asleep at the time, and there were no warnings given. Many were planning to take the day off because November 4 was the celebration of Armed Forces Day to commemorate the Italian victory over the Austrians in World War I. This was fortunate in some ways because businesses were closed and few people were in the street; however, emergency workers encountered locked doors when they needed to break in and rescue important items. Jewelers, however, were fortunate because their night watchmen had warned them, and they were able to move their merchandise.

By 7:30 a.m., gas, water, and electricity were cut off. Soon the army barracks were flooded, and the hospitals lost their emergency generators. By 9:45 a.m., the main Piazza del Duomo, home of the most famous cathedral in Florence, was flooded. The floodwaters ruptured heating tanks, turning the water into an oily, destructive mixture. In a 1969 interview, artist Marco Sassone said, "The only thing you could do was watch and be helpless. Nature was master and the women became crazy with fear. They began throwing things from the windows and screaming 'who is going to save my children?'" (Barton, *The Register*, November 2, 1969, p. 10). The city was eventually cut in two; parts were unreachable, even by boat. At its peak, the water was 6.7 m (22 feet) above the highest flood stage on the river, and nearly every building was flooded in its ground floor and basement. By 8 p.m., the worst was over, and the flood started to recede.

The flood damage and loss of life (101 people died) were made worse because Florence had so little flood preparation or control in place. The Arno rarely flooded, and there had been no example of such severe flooding since 1557. Nonetheless, 5,000 families were made homeless, and 6,000 stores had to be closed. This flood was different from any other flood in that region because of the large amount of valuable art and historical objects that were lost or damaged. About 600,000 tons of mud, rubble, and raw sewage filled basements and first floors of nearly every building. Although most works of art in the Pitti and Uffizi palaces were spared, many other historic buildings and their collections were devastated. It is estimated that the flood damaged

Fig. 5.8. A, The Ponte Santa Trinita (in the background) and the banks of the Arno River are covered with floodwaters on November 4, 1966. B, The first floors of the buildings in the Piazza San Firenze are completely inundated by a mix of oil and water. (Photos from www.florin.ms/1966flood.html)

between 3 million and 4 million books and old manuscripts, as well as about 14,000 pieces of art. The damage would have been much greater if not for the efforts of hundreds of volunteers who carried most objects to safety and worked to clean them up. They came to be known as the "mud angels" (*angeli del fango*). Even now, some 45 years after the event, thousands of books, manuscripts, and artwork have never been restored because of limited funds and staffing.

Although the damage to culture was severe in the 1966 Florence floods, relatively few people died. The world's deadliest floods all occur in Asia (especially China), where huge populations of peasant farmers live on floodplains, with limited help from governmental or international disaster relief. The all-time deadliest floods were those that struck central China in 1931, swelling all of China's rivers, including the Yellow (Huang He), Yangtze (Chang Jiang), and Huai. The region not only received record rainfall (more than 60 cm, or 2 feet, of rainfall in two months) but also seven typhoons struck China during that time. It is estimated that almost 4 million people died from the flood or starvation and from the spread of diseases afterward. Nearly 90,000 km² (34,700 square miles) were swamped, and 80,000 people were left homeless. Chiang Kai-shek's Nationalist government was too weak to deal with the devastation because it was concerned with its own civil war against the Communists as well as constant warfare with the aggressive Japanese empire. This flood was certainly the deadliest natural disaster of the twentieth century and probably one of the deadliest of all time.

The world's second deadliest flood is the 1887 Yellow River flood (2 million dead), where heavy fall rains collapsed dikes and flooded northern China, drowning more than a million; a million more died from starvation. The third worst flood in history was the 1938 Yellow River flood (700,000 dead), although this was a man-made disaster. Hard pressed by the Japanese Imperial Army, the Chinese Nationalist government and its armies opened the Huayuankou dike along the Yellow River to stop the Japanese invasion. Chinese peasants downstream were deliberately not told out for fear of alerting the Japanese to their plans. Hundreds of thousands of innocent Chinese civilians died. More than

54,000 km² (21,000 square miles) were inundated. The flood changed the channel direction of the Yellow River for several years. Some people say the move was part of war and justified militarily because it stalled the advance of the Japanese army for more than a year. However, many others consider it an atrocity committed by a government against its own people.

The fourth deadliest flood was the failure of the Banqiao Dam in China in 1975, as a result of Typhoon Nina, with 231,000 dead (discussed further in chapter 6). The fifth worst flood resulted from the Indian Ocean tsunami; however, the sixth worst was the 1935 Yangtze flood, which killed 145,000. This was China's third catastrophically deadly flood during the 1930s. Today, China has embarked on a massive program of damming rivers, not only for hydroelectric power but also to mitigate the long and sorry history of flooding disasters. The controversial Three Gorges Dam, originally proposed in 1919 by Sun Yat-sen and promoted by Chiang Kai-shek and Mao Zedong, was completed in 2008. It will permanently modify the flow of the Yangtze River but at a great cost to the 1.24 million displaced people, archaeological sites, endangered species, and natural habitats that will be drowned when the reservoir is filled.

Flooding in a Flash

Massive seasonal floods in floodplains and other regions of extensive rainfall and low elevation can be catastrophic but usually warnings can be issued because many scientists monitor the weather and rainfall amounts. In drier parts of the world, such as semideserts of the western United States, however, floods are different. Rare desert downpours are typically violent and strike with little warning. Often, these desert rains drop many inches of rain in a few hours, turning a dry riverbed into a raging torrent in a matter of minutes. These events are known as flash floods, because they strike so quickly on such a large scale with little warning. It may not be raining where you are, but a rainstorm in a distant mountain range may suddenly bring a wall of water down the canyon with no place to climb out and no place to outrun it. There are many stories of such floods trapping hikers and drivers, who find themselves unable to escape and are washed away and drowned. The beautiful slot canyons in northern Arizona are popular with tourists and photographers, but more than once, people have been trapped in them when a downpour many miles away suddenly flooded the canyon. This happened in Antelope Canyon, Arizona, a mecca for photographers. On August 12, 1997, a flash flood 3.3 m (11 feet) high rushed down the canyon and killed 12 hikers who could not climb the steep walls, while their friends watched from above, helpless.

A flash flood occurred in Big Thompson Canyon, which flows out the Colorado Rockies north of Denver and Boulder. The Big Thompson River drains most of the tributaries that come east out of Rocky Mountain National Park and Estes Park on their way down to the plains in Loveland, Colorado. In most years, the canyon is nearly dry or has a small creek tumbling through the boulders. During the spring snowmelt, the river runs high. Big Thompson Canyon is lined with campgrounds, cabins, motels, and many other features for campers, fishermen, and tourists seeking a beautiful mountain retreat.

On July 31, 1976, thousands of people were vacationing in the canyon, at the start of a three-day weekend celebrating the July 30 centennial of Colorado's statehood; the nation had just celebrated its own bicentennial. Warm winds blew moisture from the east, making towering thunderheads that rose up as they ascended the mountain front. By 7:00 that evening, the rain poured down with an intensity not even old-timers could remember. Rain gauges recorded 300 mm (12 inches) of rain in less than 4 hours, about 75 percent of the total annual rainfall for the area. The river's discharge was soon four times the maximum recorded at any time in the previous century, jumping from 137 cubic feet/sec to 31,200 cubic feet/sec in just 3 hours. The river quickly rose and formed a turbulent churning mass of water, rocks, and mud moving 25 km/h (15 mph) in a wall of water more than 6 m (20 feet) high, which could not be outrun. Most of the 400 cars on Highway 34 were turned into projectiles as they tumbled about. Some people abandoned their cars and climbed to safety on the canyon walls, although they spent a long, cold, terrifying night trapped up there. Those who stayed in their cars and tried to outrun the flood died.

The steep walls of the surrounding canyon also contributed to many landslides, which amplified the amount of debris being carried. Giant boulders, huge trees, cars, and even houses were carried downstream, wiping out all the roads and bridges as they moved (plate 8). The largest rock moved by the flood weighed 75 tons. When the flood finally subsided, cars were wrapped around trees, huge logjams clogged the canyon, and most of the bridges and roads were gone. At least 144 people died, most buried in debris from the rocks and trees, and some were never found. Four hundred and eighteen houses were destroyed, along with 52 businesses and 400 cars. Damage totaled more than $40 million in 1976 dollars. In its aftermath, zoning laws were passed to limit the building of permanent residences in dangerous canyons like Big Thompson. Today, the canyon has been restored to its former beauty, and there are only a few subtle memorials of the disaster that hit more than 34 years ago. The highway signs in the region now advise, "Climb to safety in case of flash flood!"

FOR FURTHER READING

Allen, J. E., M. Burns, and S. C. Sargent. 1986. *Cataclysms on the Columbia.* Timber Press, Portland, OR.

Alt, D. M. 2001. *Glacial Lake Missoula and its Humongous Floods.* Mountain Press, Missoula, MT.

Bretz, J H. 1923. The channeled scabland of the Columbia Plateau. *Journal of Geology* 31:617–49.

Bretz, J H., H. T. U. Smith, and G. E. Neff. 1956. Channeled scabland of Washington—new data and interpretations. *Geological Society of America Bulletin* 67:957–1049.

Changnon, S. A. 1996. *The Great Flood of 1993: Causes, Impacts, and Responses.* Westview Press, Boulder, CO.

Clark, R. 2008. *Dark Water: Flood and Redemption in a City of Masterpieces.* Doubleday, New York.

Mathur, A., and D. D. Cunha. 2001. *Mississippi Flood: Designing a Shifting Landscape.* Yale University Press, New Haven, CT.

Miller, E. W., and R. M. Miller. 2000. *Natural Disasters: Floods.* ABC-Clio, New York.

Smith, K., and R. Ward. 1998. *Floods: Physical Processes and Human Impacts.* John Wiley, New York.

Soennichsen, J. 2008. *Bretz's Flood: The Remarkable Story of a Rebel Geologist and the World's Greatest Flood.* Sasquatch Books, Seattle, WA.

Weis, P. L., and W. L. Newman. 1976. *The Channeled Scablands of Eastern Washington: The Geologic Story of the Spokane Flood*. U.S. Geological Survey, Washington, DC. www.nps.gov/history/history/online_books/geology/publications/inf/72-2/.

Hurricanes, Cyclones, and Typhoons

Nature on the Rampage

This is the disintegrating power of a great wind: it isolates one from one's own kind. An earthquake, a landslip, an avalanche, overtake a man incidentally as it were—without passion. A furious gale attacks him like a personal enemy, tries to grasp his limbs, fastens upon his mind, seeks to rout his very spirit out of him.
—Joseph Conrad, *Typhoon*, 1903

The Hurricanes of 2005

Residents of the southeastern United States, Mexico, and the Caribbean know it as "hurricane season." Officially spanning June 1 to November 1, hurricane season is the time of year when most of the hurricanes come out of the equatorial Atlantic, sweep up the Gulf Stream, and strike the eastern part of North and Central America. This is when the heat of the summer has fully warmed up the tropical waters and built up enough energy to trigger hurricanes. In some years, only a few tropical storms become powerful enough to be called hurricanes, but in the past few decades, hurricanes have been more powerful with each season.

The year 2005 was turning out to be the worst hurricane season in recorded history. By the end of the season in January 2006, there were a record 28 large, officially named tropical storms, and a record 15 had become hurricanes. Of these, seven turned into major hurricanes, a record five of them reached Category 4 on the Saffir-Simpson scale of hurricanes (extreme hurricanes with wind speeds exceeding 210 km/h (131 mph), and a record four of them reached Category 5 (catastrophic hurricanes with wind speeds exceeding 250 km/h, or 155 mph). Among these was the most intense storm ever recorded. There were even hurricanes in the South Atlantic, a phenomenon that scientists thought was impossible. There had never been a season like it, but there may be more of these in store.

The season got off to a rip-roaring start in June with tropical storms Arlene and Bret and Hurricane Cindy, which dropped 133 mm (5 inches) of rain on Louisiana and Mississippi and killed three people. On July 5, Hurricane Dennis struck Cuba, Haiti, and Florida, becoming a Category 4 storm with the strongest air pressure ever recorded in July in that region; it killed 89 people and caused $5 billion in damage. As people were recovering, Hurricane Emily closed in, setting the record for the fastest-moving storm. It hit southeastern Mexico on July 11 with Category 5 strength, killing 14 people and causing $400 million in damages. These storms alone would have made 2005 one of the worst seasons on record, but it wasn't even August yet. August 2005 turned out to be much worse than even the most pessimistic forecasts. After tropical storms Franklin, Gert, and Harvey passed without major damage, Hurricane Irene was followed by tropical storm Jose, which flooded eastern Mexico even though the storm never reached hurricane status.

Then came Katrina (fig. 6.1). Forming over the Bahamas in mid-August, it became a Category 4 hurricane as it passed over Florida, reached the warm Loop Current in the Gulf of Mexico, and built up even more energy. Meteorologists had been tracking it for days and sent out warnings to officials and the

Fig. 6.1. A NOAA satellite image of Hurricane Katrina before it made landfall in Florida, August 2005. In the next few days, it gained energy over the Gulf of Mexico and then slammed into New Orleans and the southern Gulf Coast. (Courtesy NOAA)

public that this was going to be a catastrophic storm, but neither the local authorities nor the Bush administration took the extra steps that were required: earlier and more urgent evacuations, sending relief aid as the storm approached, and so on. Local agencies warned most of the area in the storm's path to get out of the way. When Katrina made its second landfall along the southern Gulf Coast on August 29, 80 percent of the people in the region had evacuated, but others were hunkered down, figuring they could ride out the storm. The evacuation order was useless to poorer neighborhoods of New Orleans, whose residents did not have cars, and no public transportation was provided.

Katrina turned out to be much worse than anyone had anticipated (plates 9, 10). The damage from the heavy wind and rain was more severe than during any hurricane in recent memory, and buildings and trees up and down the

Gulf Coast of Alabama, Mississippi, and Louisiana were devastated. Even more catastrophic was the huge two-story-tall wall of water, or storm surge, that poured inland due to the hurricane. This water broke through dikes and levees of New Orleans in more than 50 places and released a huge volume of pollution-laden water into the low-lying districts, flooding many houses to their rooftops, and inundating 80 percent of the city.

News was filled with images of people and their pets stranded on rooftops (plate 9B), awaiting rescue by helicopters and boats. There were images of evacuees crowded into the New Orleans Superdome, Civic Center, and the Houston Astrodome without adequate food, water, or medical care. As the waters receded, images of the Lower Ninth Ward completely under water were broadcast, and many other parts of New Orleans drowned in several feet of water. Most shocking were images of corpses rotting in the sun—drowning victims that could not be identified.

So far, Katrina had caused at least $82 billion in damage, the costliest natural disaster in U.S. history, more than twice the previous record held by Hurricane Andrew in 1982 (calculated in 2008 dollars). Katrina killed at least 1,836 people (with more than 700 still missing and unaccounted for), making it the deadliest U.S. hurricane since the 1928 Lake Okeechobee hurricane. Katrina permanently transformed the city of New Orleans. Many poor residents moved outside of New Orleans because their homes were lost or damaged beyond repair. Even houses that had only seen moderate flooding were condemned because of water damage and pollutants that were mixed in, which made structures uninhabitable, especially after months of mold, mildew, and rot had set in. Katrina became a symbol of the Bush administration's bungling and insensitivity to poor people, and a warning to New Orleans that a town built largely at or below sea level would always be vulnerable.

Lost in all the political finger-pointing was a larger problem: Katrina was more of a human-induced disaster than a natural one. Katrina was a strong storm, but it had dropped to a Category 3 hurricane before landfall, not enough to top the levees, which were supposedly built to survive a storm of that size. Katrina revealed that because of political bungling and pork-barrel spending billions of dollars had been spent on less essential projects like streets and sewers but not enough on the construction and maintenance of the levees of New Orleans. Geologists who studied the aftermath identified a larger problem: years of draining and developing wetlands along the Gulf Coast had worsened hurricane danger. At one time, natural wetlands absorbed the shock of hurricanes and prevented them from spreading their damage inland. A *Time* magazine article published on the second anniversary of Katrina in 2007 revealed that little had been done to mitigate the problem. The old levees had been patched up and billions of dollars that Congress and President Bush threw at the problem were wasted on temporary patch jobs and hundreds of unused trailers without really making New Orleans safer for the next big hurricane.

Even though Katrina dominated the news in the United States for months, the 2005 hurricane season was far from over. On September 2, Hurricane Maria became a Category 3 storm and then turned into a major extratropical hurricane, eventually reaching Iceland and Norway, where it caused death and destruction. Next were Hurricanes Nate and Ophelia, which slammed into North Carolina and rolled up the coast, making landfall again in Maritime Canada, causing $70 million in damages. The real kicker was Hurricane Rita, which hit the Florida Keys and became a Category 5 on September 21 (third largest hurricane on record) as it struck the Texas-Louisiana border region. Rita destroyed much of the oil infrastructure in the region and devastated the bayous of western Louisiana, killing at least 113 and causing $10 billion in damage. TV news was filled with images of thousands of people leaving Houston, Texas (fourth largest city in the U.S.), clogging the roads during the evacuation. Houston got lucky and was not hit as badly as was predicted.

By October, the Atlantic hurricane season normally winds down but not so in 2005. On October 2, Hurricane Stan struck the east coast of Mexico (already damaged from earlier hurricanes that summer) and caused intense flooding and mudslides, with more than 1,000 dead. Hurricane Tammy hit Florida on October 5, and eventually caused heavy rainfall and flooding on the northeast Atlantic Coast. Hurricane Vince struck Spain a few days later, the first hurricane in that region since 1842. Finally, on October 17 came the grand finale: Hurricane Wilma. It was the strongest hurricane on record in the Atlantic, with wind speeds of 185 km/h (255 mph) and pressures of 882 millibars, one of the highest pressures ever measured. First, it struck Cancun and Cozumel, becoming the most damaging storm in Mexican history, before curving back to pound Cuba and Florida. It caused 23 deaths and more than $29 billion in damages.

The standard system for naming Atlantic hurricanes uses 21 names in alphabetical order (see table 6.1), but the 2005 hurricane season was still not over. Hurricane Wilma constituted the 21st and last named hurricane. Afterward, Greek letters were used for the first time. Tropical storm Alpha flooded Haiti and the Dominican Republic and killed 42 people. Hurricane Beta hit Nicaragua on October 26 with Category 3 strength. Tropical storm Gamma killed 41 people in Honduras and Belize on November 15. Tropical storm Delta clobbered the Canary Islands on November 28, causing severe damage and many deaths. The final two storms, Hurricane Epsilon and Tropical Storm Zeta, were not as deadly, but Zeta became the latest tropical storm on record, finally dissipating on January 6, 2006.

After the 2005 season, the World Meteorological Organization retired five entries from their rotating list of tropical storm names (table 6.1). Future tropical storms or hurricanes will never be named Dennis, Katrina, Rita, Stan, and Wilma. Table 6.2 lists retired names, which are permanently associated in the memories of victims with particular catastrophic storms. When the old 2005 name list cycles back in 2011, the retired names will be replaced by Don, Katia, Rina, Sean, and Whitney. In the long tradition of male-dominated fields, where ships and hurricanes were named after women, the list originally had only female names. However, in recent years, the official list alternates between male and female names, using names from around the world, not only Anglo names. The letters Q, U, or X, Y, and Z are excluded. The compilation committee had enough challenges finding names for the less commonly used letters of the alphabet, let alone trying to find six or more male and female names that began with Q or X (table 6.2).

Table 6.1. Official alphabetical list of hurricane names
(names repeat in six-year rotation)

2007	2008	2009	2010	2011	2012
Andrea	Arthur	Ana	Alex	Arlene	Alberto
Barry	Bertha	Bill	Bonnie	Bret	Beryl
Chantal	Cristobal	Claudette	Colin	Cindy	Chris
Dean	Dolly	Danny	Danielle	Don	Debby
Erin	Edouard	Erika	Earl	Emily	Ernesto
Felix	Fay	Fred	Fiona	Franklin	Florence
Gabrielle	Gustav	Grace	Gaston	Gert	Gordon
Humberto	Hanna	Henri	Hermine	Harvey	Helene
Ingrid	Ike	Ida	Igor	Irene	Isaac
Jerry	Josephine	Joaquin	Julia	Jose	Joyce
Karen	Kyle	Kate	Karl	Katia	Kirk
Lorenzo	Laura	Larry	Lisa	Lee	Leslie
Melissa	Marco	Mindy	Matthew	Maria	Michael
Noel	Nana	Nicholas	Nicole	Nate	Nadine
Olga	Omar	Odette	Otto	Ophelia	Oscar
Pablo	Paloma	Peter	Paula	Philippe	Patty
Rebekah	Rene	Rose	Richard	Rina	Rafael
Sebastien	Sally	Sam	Shary	Sean	Sandy
Tanya	Teddy	Teresa	Tomas	Tammy	Tony
Van	Vicky	Victor	Virginie	Vince	Valerie
Wendy	Wilfred	Wanda	Walter	Whitney	William

Scientists worry that the 2005 season was part of a long-term trend toward more intense hurricanes over the past 20 years. That season broke the 2004 record, the previous worst season on record. The number of Category 4 and 5 storms between 1990 and 2006 increased dramatically compared with the interval between 1975 and 1989 (Curry et al. 2006). Emanuel (2005) showed that the power of the most recent hurricanes (as measured by wind speed and duration) had increased 50 percent since 1970. Nearly all the most destructive hurricane seasons on record had occurred in the past 20 years, with many of

Table 6.2. Official list of retired hurricane names

Agnes, 1972	Donna, 1960	Iris, 2001
Alicia, 1983	Dora, 1964	Isabel, 2003
Allen, 1980	Edna, 1968	Isidore, 2002
Allison, 2001	Elena, 1985	Ivan, 2004
Andrew, 1992	Eloise, 1975	Janet, 1955
Anita, 1977	Fabian, 2003	Jeanne, 2004
Audrey, 1957	Fifi, 1974	Joan, 1988
Betsy, 1965	Flora, 1963	Juan, 2003
Beulah, 1967	Floyd, 1999	Katrina, 2005
Bob, 1991	Fran, 1996	Keith, 2000
Camille, 1969	Frances, 2004	Klaus, 1990
Carla, 1961	Frederic, 1979	Lenny, 1999
Carmen, 1974	Georges, 1998	Lili, 2002
Carol, 1954	Gilbert, 1988	Luis, 1995
Celia, 1970	Gloria, 1985	Marilyn, 1995
Cesar, 1996	Hattie, 1961	Michelle, 2001
Charley, 2004	Hazel, 1954	Mitch, 1998
Cleo, 1964	Hilda, 1964	Opal, 1995
Connie, 1955	Hortense, 1996	Rita, 2005
David, 1979	Hugo, 1989	Roxanne, 1995
Dennis, 2005	Inez, 1966	Stan, 2005
Diana, 1990	Ione, 1955	Wilma, 2005
Diane, 1955		

Note: Influential hurricanes and catastrophic storms have their names retired.

the years in the 2000s at or near record levels of major hurricanes. The ranking of seasons with the most major hurricanes has 2005, 1999, 1996, and 1994 in the top five. Of the seasons with the most named tropical storms in the Atlantic, 1995, 2000, 2001, 2003, 2004, 2005, 2007, and 2008 are all on the top ten list, and only two years (1933, 1936) on the top ten list were before 1969. Only 2002 and 2006 have failed to make the list of the most named tropical storms in this decade. The 2009 season started late, but it had a record 7 named storms in August, and finished with 20 storms and 8 hurricanes altogether. The year 2010 was the most active season since 2005, with 19 named storms and 12 hurricanes.

One criticism of efforts to identify whether hurricanes have become worse in recent years is the lack of a detailed historical record. Other than a few great storms of past centuries, there are almost no reliable historical records of hurricanes before 1900. A study by Mann and colleagues (2009) used cores taken from seven coastal locations (lagoons, lakes, barrier islands) on the Atlantic Coast of the United States and one site in Puerto Rico and identified the sandy storm layers, dating them by radiocarbon methods. They obtained a historical record of almost a thousand years of hurricane seasons on the Atlantic. They found that the past two decades have had the highest number of hurricanes, averaging 17 per year, more than twice the value of any previous decade. The last time hurricanes were this frequent was during the Medieval Warm Period.

Not surprisingly, this information has fueled the debate over climate change. That the sea surface temperatures in the tropical oceans have warmed dramatically cannot be disputed, even among climate skeptics. Warmer oceans, which generate hurricanes in the first place, produce more hurricanes and intense storms. Many top scientists argue that the connection between hurricanes and global warming is real (e.g., Curry et al. 2006; Emanuel 2005b; Webster et al. 2005), and most of these articles have been published in stringently reviewed, highly regarded scientific journals such as *Science* and *Nature*. However, climate skeptics are not convinced. The data for both oceanic temperatures and hurricane velocities do not go back far enough, so it is hard to establish past correlations.

Galveston, Oh Galveston!

Although Hurricane Katrina is deeply etched in the American memory, the deadliest hurricane to strike the United States occurred more than a century ago. It is also the deadliest of any kind of natural disaster (including earthquakes, volcanoes, and landslides) to strike in the United States. It is only known from archival records because there are no survivors alive today from the storm that leveled Galveston, Texas, in 1900. At the turn of the twentieth century, Galveston was a boomtown, the "Jewel of Texas" and the home of Texas's first post office, telephones, and medical college. It was a busy port town as well as a financial center, known as the "Wall Street of the Southwest." Ships brought in goods from all over the world and exported tons of cotton from the Deep South. It was the largest city in Texas, with a population exceeding 43,000, and was still growing and developing.

In the summer of 1900, a storm developed 4,000 miles away, far across the Atlantic near the Cape Verde Islands off the coast of West Africa. Today, weather satellites, meteorological aircraft, and weather stations monitor every stage of a tropical storm; however, in 1900, there was no way to detect the storm in the open Atlantic except for the casual anecdotes of ship captains at sea. The first record was on August 27, 1900, when a ship captain mentioned "unsettled weather" about 1,600 km (1,000 miles) east of the Leeward Islands in the Caribbean. By August 30, the storm had reached the Windward Islands and Antigua, where it was still a tropical depression with severe thunderstorms. By September 1, the U.S. Weather Bureau reported a "storm of moderate intensity (not a hurricane)" passing south of Cuba, which was drenched by the storm on September 3. By September 6, the storm was north of Key West, Florida, and on September 7, weather observers reported a severe storm had hit the Gulf Coast of Louisiana with heavy damage, but they were unable to pass the word to other areas because the storm had wiped out communications. The Galveston Weather Bureau issued hurricane warnings by the afternoon of September 7, and the ship *Louisiana* (which sailed from Galveston that morning) signaled that it was experiencing winds of 160 km/h (100 mph) as it passed through the storm. This would make the hurricane a Category 2 on the Saffir-Simpson scale, and it was still picking up heat and moisture from the Gulf and had not yet made landfall.

Even with all the warnings, however, most residents of Galveston were not alarmed. Galveston was built on a low, sandy barrier island that protected the Trinity River's Galveston Bay, which flowed through Houston. At the time, its highest point was only 2.7 m (9 feet) above sea level, no match for storm waves of more than 4.6 m (15 feet) tall that would sweep across the island. Geographers assured Galveston residents that the long, gentle gradient of the continental shelf offshore would buffer the effects of the storm, dissipating its energy by the time it made landfall. A seawall to protect the island was proposed, but no one was concerned enough to build it. Residents were seduced by the myth that any future storm would be no worse than the previous storms they had experienced. In an 1891 article in the *Galveston Daily News*, the Galveston Weather Bureau section director Isaac Cline argued that a seawall was *not* needed to protect the city and maintained that it was impossible for a large hurricane to strike the island. Development of the island made it even more vulnerable. A large number of flimsy wooden houses were built in the sand, and most of the coastal sand dunes had been excavated as fill dirt for construction, which further eroded the island's natural barriers.

Ironically, longtime Texas residents should have known better. Indianola, a nearby town on the Matagorda Bay, was built on a similar barrier island and was nearly wiped out during a hurricane in 1875. When a second, even more powerful hurricane struck in 1886, the town was destroyed, and survivors gave up trying to rebuild and moved elsewhere. Galveston residents did not heed the lessons of 14 years earlier, believing that their situation was different from Indianola's.

As the storm approached, fewer than half of the island's residents evacuated to the mainland, and some Houstonians actually came to Galveston to gawk at the storm. Early on the morning of September 8, a huge storm was brewing, and an evacuation order was given. It was too late because the waters of Galveston Bay were too rough for most boats to cross, and some ferries had broken loose from their moorings. One steamship floated away, out of control, and broke the three bridges that connected Galveston Island to the mainland, stranding the remaining residents.

Throughout Saturday, September 8, the Category 4 storm struck with maximum wind speeds of 240 km/h (150 mph). The barometric pressure plunged to such a low reading that the observers thought the barometer was broken; they had never experienced a low-pressure cell this intense. The huge storm surge 4.6 m (15 feet) tall formed a wall of water that knocked most of the buildings off their foundations and reduced the wooden ones to splinters. Brick and stone buildings remained but were largely ruined as well because the storm waves pounded them with debris from buildings and ships like battering rams. The rest of the town was turned into a pile of broken wood and rubble (fig. 6.2). The official death toll was 8,000, although most sources place it closer to 12,000. More people were killed in this natural disaster than in any other in U.S. history. Most were killed when they drowned in the storm surge or were crushed when buildings collapsed and waves pounded the debris around them. Many more died in the next few days because of injuries that could not be treated. Others were trapped in the wreckage and died of thirst and starvation. Damage was about $20 million in 1900 dollars, or $516 million in 2009 dollars.

By Sunday morning, September 9, clear skies and mild winds greeted survivors, who walked around in shock amid the rubble and attempted to pick through the wreckage to find survivors still trapped or to rescue their valuables. Meanwhile, the storm continued inland, where it pounded Texas and Oklahoma and reached Milwaukee with sustained winds of 40 mph. In New York City, sustained winds of 105 km/h (65 mph) tore down signs and awnings and killed several people. When it reached Halifax, Nova Scotia, on September 12, it clobbered the fishing fleet off Newfoundland before fading in the North Atlantic.

Galveston Island was isolated and lacked communications to the outside world; relief efforts to the island were difficult. A train leaving Houston for Galveston as the storm began on September 8 was forced to stop when it found the tracks washed out and the ferries gone. Some train passengers sought shelter in a Bolivar Lighthouse and survived, crammed like sardines amid the stench and smell of human excrement as 10 m (33 foot) waves pounded the walls of the lighthouse. But all those who stayed in the train drowned when the storm surge overran the cars.

On Monday, September 9, the ship *Pherabe*, one of the few ships originally docked in Galveston that had survived the storm, arrived in Texas City on the

Fig. 6.2. Pictures of the devastation of Galveston after the 1900 hurricane. A, House blown over on its side. B, Ship pushed out into the wreckage of the city. C, Galveston was totally flattened into piles of lumber; a body lies in the wreckage of the wharf. (Images from old lantern slides; courtesy Wikimedia Commons)

mainland and sent out messages, which soon spread to Houston and other cities. In a telegram, the governor of Texas, Joe Sayers, sent President William McKinley the following news: "I have been deputized by the mayor and Citizen's Committee of Galveston to inform you that the city of Galveston is in ruins" (Green 1900, 126). G. L. Vaughn, the Western Union manager in Houston, sent the following message to the head of the U.S. Weather Bureau:

> First news from Galveston just received by train which could get no closer to the bay shore than six miles where Prairie was strewn with debris and dead bodies. About 200 corpses counted from train. Large Steamship stranded two miles inland. Nothing could be seen of Galveston. Loss of life and property undoubtedly most appalling. Weather clear and bright here with gentle southeast wind. (Green 1900, 126)

Trains and boats from Houston and elsewhere soon rushed to the island to help survivors. Early messages estimated the death toll at only 500, and even that seemed incredible to outsiders. Little did they know that at least ten to fifteen times that many were actually dead. Almost a quarter of the population had died in one day, and there were not enough places to store the dead bodies (fig. 6.2C). It was impossible to bury the hundreds of bodies. When corpses were dropped in the ocean, they floated back to shore. Therefore, huge funeral pyres were set up to burn bodies for weeks to prevent the stench and to stop the spread of disease from the rotting corpses. The authorities passed out free whiskey to the men who had to throw hundreds of bodies, some of them their own wives and children, onto the bonfire.

Within days, the railroad lines and ferries were restored, mail and telegraphs were working again, and fresh water was accessible. A huge tent city, the "White City on the Beach," was constructed from U.S. army tents. Many people built temporary houses out of "storm lumber" salvaged from the ruins. Eventually, bodies were all burned and the wreckage was cleared, but the storm changed many residents' minds about living on Galveston Island. Development soon shifted to Houston, which was beginning to experience its first oil boom. Galveston was eventually rebuilt, with sand dredged from the island, raising the city higher above sea level, and a huge seawall was built to protect it. Galveston never recovered its status as the largest city in Texas. A hurricane in 1915 claimed 275 lives. Life on a barrier island is never safe from hurricanes.

Hurricanes, Typhoons, and Cyclones

The terminology of great tropical storms can be confusing. By convention, tropical storms that arise in the Atlantic and eastern Pacific oceans become *hurricanes*, those that occur in the western Pacific Ocean are known as *typhoons*, and those that originate in the Indian Ocean are known as *cyclones* (fig. 6.3). They occur only in a belt just above and below the equator in these oceans, but not at the equator, since the prevailing winds do not form a strong spiraling pattern that promotes the spiral ("cyclonic") flow of hurricanes. All tropical storms arise from similar forces: excess heat in the ocean. Heat that has built up in these tropical oceans over the long summer must be dissipated to higher latitudes because heat always flows from hotter to colder regions. Once the sea surface temperature has reached a threshold of 27°C (80°F) in the upper 60 m (200 feet) of water, conditions are right for a tropical storm. If the air above the ocean is also warm, humid, and unstable, and there are no strong upper-level winds to move the heat away, a tropical storm develops, typically when there is a large area of low pressure over the tropical Atlantic, Pacific, or Indian oceans.

This low-pressure zone begins with a series of small, disorganized thunderstorms with weak surface winds, known as a *tropical disturbance*. If the surface winds become stronger and better organized, they soon spiral around the area of low pressure in a cyclonic fashion around a central core, forming a *tropical depression*. These winds then converge on the spiraling area, with the core acting as a chimney, allowing warm moist air to rise quickly to the stratosphere and sucks more moisture out of the clouds that spiral around it. The rising moist air quickly cools and water condenses, releasing huge amounts of latent heat. This in turn warms the surrounding air and causes stronger updrafts,

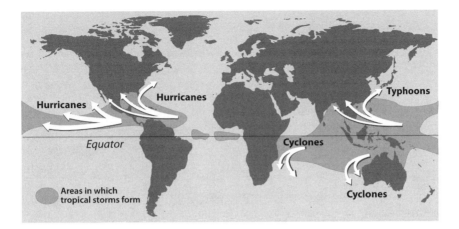

Fig. 6.3. Map where tropical storms occur. (Courtesy USGS; redrawn by Pat Linse)

which in turn increase the rate of flow of warm moist air from below, creating a feedback loop that amplifies the storm until it reaches a massive size.

As the converging winds spiral around the low-pressure center, they increase in strength. When winds exceed 63 km/h (39 mph), they have become a *tropical storm*. The system can grow even larger if it is fed by warm, moist waters and continues to grow in size and power, with higher and higher wind speeds. If it exceeds wind velocities of 119 km/h (79 mph), it then becomes a hurricane, a typhoon, or a cyclone. On the Saffir-Simpson scale, there are five categories of hurricanes, from Category 1 (wind speeds of 119–153 km/h, or 74–95 mph) up to Category 4 and 5, mentioned earlier.

Hurricanes are basically heat engines to transfer energy from the tropics to higher latitudes. An average hurricane generates energy at rates 200 times as fast as our worldwide capacity to generate electricity. The kinetic energy of the winds in a single hurricane is about half the worldwide global energy capacity, and the energy released in a hurricane through its clouds and rain is 400 times as powerful as the energy of its winds.

Once a hurricane develops, it becomes the huge spiraling mass of clouds familiar on satellite images (figs. 6.1, 6.4). Most hurricanes are made of a series of cyclonic rain bands that spiral counterclockwise around the eye and inward,

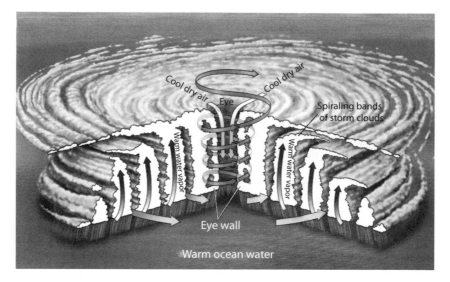

Fig. 6.4. Diagram of a hurricane, showing the structure and wind direction. (Courtesy USGS; redrawn by Pat Linse)

pulling warm, moist air from the ocean. In the center is the eye of the hurricane, where the wind spirals up the "wall" of the eye in an intense updraft and cool air from above sinks rapidly down the center of the eye to counter the upward flow (fig. 6.4). They are then moved forward by the trade winds and their own energy, forming a curved path (fig. 6.3). In the Atlantic, the paths all move the same way, in a clockwise, recurved path toward the northeast and then out to the northwest, because the trade winds spiral around a large high-pressure cell in the north-central Atlantic called the Bermuda high. As long as these systems stay over warm water, they can increase in energy and develop higher wind speeds. Once they move over cold water or land, the energy is quickly dissipated, and the system breaks down and dies out.

Monster Typhoons

Although U.S. residents don't pay as much attention to them, the typhoons of the western Pacific have recorded winds stronger than the strongest Atlantic hurricane. The Pacific also has the most intense storms on record anywhere in the world. They typically result in huge loss of life when they flood low-lying regions in Asia where hundreds of poor farmers and peasants live. Their damage to structures can be even more severe because they hit poorly constructed buildings in relatively poor countries. The name *typhoon* comes from Cantonese *dai fung* or Mandarin *da feng*, which means "big wind" in both dialects. Although they form year round, typhoons are most common from August to October, because the same peak of late summer and fall oceanic warming that drives the Atlantic hurricane season also produces typhoons. They originate in the western equatorial Pacific, then (like Atlantic hurricanes) trend northeast before curving back to the northwest (fig. 6.3), because they are driven by prevailing winds that wrap around a long-term high-pressure cell in the north-central Pacific. This means that they typically strike China, Taiwan, Japan, Korea, the smaller islands of the west Pacific, and especially the Philippines, which has been hit by more typhoons than any other nation. The list of names of typhoons was agreed to by the 17 Asian nations that experience them, with a five-year rotation before they repeat. Unlike the Atlantic hurricanes, which are all named after people, most typhoons are named after animals, flowers, astrological signs, and a few personal names. The Japanese, however, do not use the naming system and refer to typhoons by number in order of occurrence. Tropical storms in the Atlantic are similarly numbered and do not merit names unless they reach the threshold for a category 1 hurricane on the Saffir-Simpson scale.

Nina, a supertyphoon and one of the deadliest typhoons ever recorded, started out as a tropical depression in the north-central Pacific on July 30, 1975. A subtropical ridge of high pressure kept it from curving north to the Philippines, and instead, it went on almost a straight northwest-trending path toward Taiwan and China. On August 1, weather observers noticed a rapid, huge drop in barometric pressure and an increase in wind speed, from 120 km/h (75 mph) to as high as 250 km/h (155 mph), making it a Category 5 storm. As it approached Taiwan, it weakened slightly but still hit the island hard, although the damage was mostly in the island's mountainous highlands. Then it skipped on to the mainland, coming ashore on the southeast coast of China in Fujian

Province, before curving to the right (north) and hitting Hunan Province. As it did so, it set a record for rainfall in the region, dropping a meter of rain in less than 24 hours (more than a year's typical rainfall), and flooding nearly all of coastal China. As it headed north toward Henan Province on August 5, it flooded the drainage of Huai River and caused the failure of Banqiao Dam, which collapsed and helped trigger the failure of smaller dams downstream. Meteorologists calculate that this was not a 500-year flood, but a 1-in-2,000-year flood. As the storm raged during the night of August 7, there were communication problems. No one told the workers to open the dam and release the water before the catastrophic failure. By August 8, 62 smaller dams downstream were failing, and by 1:00 a.m. the next day, Banqiao Dam collapsed. It released 78,800 m³/sec (64 acre feet/second) in a huge wall of water weighing 701 million tons, moving through the valley over a 6-hour span. Altogether, the combined dams released 15.7 billion tons of water in a few hours. The wall of water was 10 km (6.2 miles) wide, 7 m (23 feet) high, and traveled at 55 km/h (31 mph), making it impossible for most people below the dams to escape. Hundreds of thousands of acres were flooded, many cities were wiped away, and at least 6 million buildings were destroyed. The severity of this disaster was unknown to the Western world because China was still a closed country in 1975. The government strictly censored most news (especially bad news) from reaching the world or their own people. It was not until 2005 that details were declassified and westerners learned about the extent of the disaster.

Altogether, the flooding from Typhoon Nina and its side effects killed at least 170,000 people—26,000 from drowning, and 146,000 from disease and famine afterward. Nina caused at least $5 billion in damage in 2009 dollars, making it one of the most expensive natural disasters ever, as well as the second deadliest typhoon in history. The record is held by an 1881 typhoon that hit Haiphong in northern Vietnam, killing at least 300,000 people. Unfortunately, little is known about this record-breaking event because it occurred more than a century ago, and there were no records or meteorological data comparable to storms we have studied in the past 50 years.

A Killer Named Nargis

If the deadliest storms occur in the western Pacific typhoon belt, a close second would be the Indian Ocean tropical cyclones. These monster storms often strike low-lying poorer regions of such countries as India, Burma, Bangladesh, and Thailand, and drown thousands of peasant farmers who have no place to escape. The deadliest named storm in all of history was Typhoon Nina, but the second deadliest was Cyclone Nargis in 2008. Nargis is the Urdu name for "daffodil" or "flower" and is a common girl's name in parts of southern Asia. This Nargis was no flower or girl but a disaster that killed hundreds of thousands, and brought Burma to its knees.

Nargis started as a strong tropical depression that developed in the center of the Indian Ocean on April 27, 2008. By April 28, it had stalled and was building up energy, reaching Category 1 status. At first, it appeared to be headed north for India or Bangladesh, but on May 1, the storm intensified and headed northeastward instead. Eastern India, Bangladesh, and Sri Lanka were not in the storm's direct path, but they experienced severe flooding from the edge of the storm and a great loss of life and damage. On May 2, Nargis reached the coast of Burma (Myanmar) as a Category 4 storm, with wind speeds clocked at 215 km/h (135 mph). As it went ashore, it brought the waves of a huge storm surge over the low, swampy Irawaddy River delta region, where millions of Burmese eke out a living growing rice. By the time it passed the capital of Yangon (Rangoon), it had slowed to 130 km/h (80 mph), but it still did massive damage to the city. Then it moved inland to the Thai-Burmese border, where it lost strength and fell apart after May 3.

The Burmese living on the Irawaddy delta had no chance. The winds and rain destroyed any shelter they sought, and the storm surge wave drowned people by the thousands and washed many out to sea. The military junta that ruled Burma refused to count the dead and gave lowball estimates of 80,000 deaths to minimize international criticism of their lack of relief efforts. This number was clearly fabricated because 80,000 died in the Labutta Township alone and another 10,000 died in Bogale. Most sources estimate the total number of dead and missing at close to 300,000, and some estimates were as high as 1 million. The true number is unknown because many bodies washed away or were buried. Besides, Burma is a closed society with no press scrutiny or access to sources of information. Another 2–3 million people were left homeless.

Once the storm had passed, it was clear from the few press accounts that got past the government censors that the scale of devastation and misery was unimaginable (fig. 6.5). At least 95 percent of the buildings in the Irawaddy delta were destroyed, as well as more than 1,400 temples. The storm surge not only wiped out the buildings and drowned millions of people but also destroyed most of the crops, because sewage from ruptured pipes washed across the fields and made the surviving rice and other food plants inedible.

More appalling was how the tyrannical regime used the disaster to destroy political opposition in the region. At first, they refused any aid from foreign countries, even though people were dying from the hurricane, from lack of clean food and water, or from rapidly spreading diseases. The government kept the world's news agencies away from the devastation and repeatedly denied its seriousness, to make their regime seem less incompetent. Foreign aid planes and ships were immediately mobilized and waited days for permission to land or dock. Still the military junta refused permission on the grounds that these foreigners were violating their sovereignty. Finally, on May 7, 2008, the Burmese government allowed the United Nations to authorize relief efforts, and on May 8, the first aid from India was allowed into the country—more than a week after the hurricane had struck. Relief efforts were still slow and cumbersome because the Burmese junta hampered travel. Visas were denied and reportedly a large share of the relief supplies were stolen to enrich the government. By late May and early June, a much larger volume of supplies reached the disaster zone but it was already too late for much of the country. Only the hardiest survivors remained; other survivors had finally succumbed to disease or starvation.

The cruelty and incompetence of the Burmese junta will probably go down

Fig. 6.5. Images from Cyclone Nargis. A, Satellite image of the cyclone as it headed east over the Bay of Bengal. (Image courtesy NOAA) B, The cyclone at full strength in Yangon (Rangoon), blowing trees apart and destroying buildings. C, A Buddha statue is all that stands among the ruins. (B and C courtesy Wikimedia Commons)

as one of the greatest acts of tyranny carried out by rulers against their people. It compares to the genocide by the Pol Pot regime in Cambodia in the 1970s or the massacres in Darfur in Ethiopia.

Damage to Burma from Nargis was staggering. Yet Bangladesh always suffers the worst loss of life. Seven of the nine more deadly weather events ever recorded have hit this country. It is a densely populated region built on the low delta deposits of the Ganges and Brahmaputra rivers, with no natural barriers to prevent flooding. More than 35 percent of Bangladesh is no higher than 6 m (20 feet) in elevation, so any rise in sea level, caused by a large storm or flood, washes across this vast low-lying area and can cause unspeakable misery. Its large (131 million), desperately poor, and rapidly growing population tries to eke three crops of rice out of the flooded fields each year, earning a household annual income of about $500. When a storm hits, 20 percent to 60 percent of the country goes under water. When cyclones hit, storm surges are 6 m (20 feet) high and drown more than 35 percent of the country. Compounding the problem is that the coastline is like a funnel, catching the cyclones and diverting them over the warm waters of the Bay of Bengal. In a typical year, Bangladesh has five cyclones in April–May (before the monsoon season) and again in October–November (as the monsoon season ends).

The Bhola cyclone of November 12–13, 1970, was a Category 3 storm that swept over East Pakistan during high tide and built a storm surge 7 m (23 feet) tall, with wind speeds reaching 255 km/h (155 mph). The wall of water drowned 500,000 people and most of their farm animals, making it the deadliest Indian Ocean cyclone of all time. The damages reached $500 million in 2009 U.S. dollars. George Harrison, formerly of the Beatles, organized the Concert for Bangladesh, which occurred on August 1, 1971, to raise money for disaster relief. It was the first in what has now become a long succession of celebrity rock concerts to raise funds and public awareness for the less fortunate. Images of starving children orphaned by the storm were used to promote the concert, which raised more than $243,000 (in 1971 dollars) for disaster relief.

The storm had an even bigger political effect. After World War II, India finally won independence from the British Empire. It became a Hindu nation, and its Muslims were sent (or migrated) to two regions: West Pakistan (what is now Pakistan) and East Pakistan on the Ganges-Brahmaputra delta. For years, East Pakistan had been trying to break away from West Pakistan. The clumsy handling of the Bhola cyclone disaster by the Pakistani government caused tremendous worldwide criticism and energized the separatist movement in East Pakistan. A year after the cyclone, East Pakistan won its independence from West Pakistan and became Bangladesh. Great storms do have consequences: not only do they kill people and force cities to be abandoned but sometimes they change the political landscape. If only the Nargis storm had done for Burma what Bhola did for Bangladesh!

A Perspective on Great Storms

As we review case histories of famous storms, patterns may emerge. First is the inherent foolishness of people who refuse to evacuate before impending disaster, especially when they have been fully warned. For events before the twentieth century, it was impossible to gauge a storm's size or to warn people adequately. Nowadays, satellites, meteorological aircraft, and ground observers closely watch every storm, so we usually know what to expect. Stubborn types insist on hunkering down through a storm, and gawkers and thrill-seekers want to experience a monster storm firsthand but often die in the process. There always seem to be incompetent officials who fail to send out proper warning or are unprepared to handle the disaster relief efforts after the storm. There are governments, such as those in China or Burma, that have a sorry history of covering up their incompetence and failing to serve their people. These unfortunate examples of human behavior result in far greater loss of life.

The larger question, however, is the wisdom of countries and communities that build on storm-prone coastlines and encourage growth on barrier islands, deltas, low-lying coasts, swamps, and other regions likely to be hit sooner or later. This is especially true of the runaway growth on the coastal barrier islands of Texas and the Carolinas, or over much of coastal Florida. People like to live near the beach, and don't think of long-term consequences or inherent risks in living in storm-prone areas. As geologists who work in these areas know, all you need to do is dig a trench down through any barrier island and layers of storm deposit are visible. Virtually an entire barrier island consists of stacked storm deposits, but almost no deposits form during the years of fair weather between storms. These risks are no secret. They are mentioned in most introductory geology textbooks. Yet people are still willing to gamble on the significant risk that they will be hurt in a natural disaster, all for the joy of living close to the beach before the next hurricane strikes.

If governmental officials allow runaway development of coastal regions and especially barrier islands, who is liable when an evacuation order results in gridlock? Hurricane prediction is much more precise than earthquake prediction, but it still has its own uncertainties. Even the best forecasts can give only a probabilistic estimate of how strong a storm will be and where exactly it will make landfall. Many storms have surprised weather watchers by suddenly changing course or by strengthening. Yet people have a short fuse for false alarms and seem intolerant when nature surprises the experts. Sometimes they even sue when a storm fails to hit as predicted because of the time and inconvenience of evacuating unnecessarily. As with earthquakes, disaster prediction in a litigious society is a no-win situation: scientists are damned if they predict the storm but people don't heed your warning, and they are damned if they don't and people sue because of the false alarm.

If people ignore warnings and build in dangerous areas, who should pay when disaster strikes? Do property owners deserve to be bailed out by private insurance companies? Do they deserve taxpayer-paid disaster relief? What happens when the impending global rise in sea level drowns these coastal communities? What happens when global warming triggers frequent and more violent storms? These issues swirl around every major disaster, but great tropical storms bring them to the fore more than any other hazard.

FOR FURTHER READING

Center for Public Integrity. 2007. *City Adrift: New Orleans Before and After Katrina.* Louisiana State University Press, Baton Rouge.

Cyson, M. E. 2006. *Come Hell or High Water: Hurricane Katrina and the Color of Disaster*. Perseus Books Group, New York.

Davies, P. 2000. *Inside the Hurricanes*. Holt, New York.

Emanuel, K. 2005a. *Divine Wind: The History and Science of Hurricanes*. Oxford University Press, Oxford.

Emanuel, K. 2005b. Increasing destrictiveness of tropical cyclones over the past 30 years. *Nature* 436:686–88.

Larson, E. 2001. *Isaac's Storm: A Man, a Time, and the Deadliest Hurricane in History*. Crown Random-House, New York.

Norcross, B. 2007. *Hurricane Almanac: The Essential Guide to Storms Past, Present, and Future*. St. Martin's Griffin, New York.

Spielman, D. G. 2007. *Katrinaville Chronicles: Images and Observations from a New Orleans Photographer*. Louisiana State University Press, Baton Rouge.

Webster, P. J., et al. 2005. Changes in tropical cyclone number, duration, and intensity in a warming environment. *Science* 309 (5742): 1844–46.

Williams, J. 2001. *Hurricane Watch: Forecasting the Deadliest Storms on Earth*. Vintage, New York.

Tornadoes

Funnels of Death

From the far north they heard a low wail of the wind, and Uncle Henry and
Dorothy could see where the long grass bowed in waves before the coming
storm. There now came a sharp whistling in the air from the south,
and as they turned their eyes that way they saw ripples in the grass
coming from that direction also.
Suddenly Uncle Henry stood up. "There's a cyclone coming, Em,"
he called to his wife. "I'll go look after the stock." . . . Aunt Em dropped her work
and came to the door. One glance told her of the danger close at hand.
"Quick, Dorothy!" she screamed. "Run for the cellar!"
—L. Frank Baum, *The Wizard of Oz*, 1900

"Tornado Alley"

Movies like *The Wizard of Oz* and *Twister* and national TV weather forecasts
have influenced our mental image of tornadoes as associated with the wide-
open plains, especially Kansas and Oklahoma. Tornadoes (fig. 7.1) are known
to occur in all fifty states, but the central Great Plains of North America have

Fig. 7.1. One of many tornadoes during the May 3, 1999, Oklahoma, tornado storm, near Anadarko, Oklahoma, that immediately preceded the F3 tornado that hit Moore, Oklahoma. (Courtesy NOAA)

a well-deserved reputation as "tornado alley" (plate 11A). More tornadoes strike the plains states (from northern Texas, through Oklahoma, Kansas, Nebraska, the Dakotas, Missouri, and Iowa) than strike any other part of the United States. The heart of tornado alley is Oklahoma. Every spring and summer, Oklahoma typically has more tornadoes than any other state or any other part of the world. The few days from May 3 to 6, 1999, stunned even veteran tornado watchers and made worldwide news. It was the most prolific outbreak of tornadoes in Oklahoma history, comparable to the Palm Sunday tornado outbreak of 1965.

May 3, 1999, started out sunny and warm. Weather forecasters predicted only a "slight risk" of tornadoes. By late afternoon, however, big thunderstorms had moved over the region, and the Storm Prediction Center in Norman, Oklahoma, had issued warnings of "high risk" of tornadoes in the ensuing hours. Later that evening, tornadoes broke out all over the state, and 66 were

Table 7.1. Fujita Tornado Damage Scale, developed in 1971
by T. Theodore Fujita of the University of Chicago

Scale	Wind Estimate (mph)	Typical Damage
F0	<73	*Light damage*. Some damage to chimneys; branches broken off trees; shallow-rooted trees pushed over; sign boards damaged.
F1	73–112	*Moderate damage*. Peels surface off roofs; mobile homes pushed off foundations or overturned; moving autos blown off roads.
F2	113–157	*Considerable damage*. Roofs torn off frame houses; mobile homes demolished; boxcars overturned; large trees snapped or uprooted; light-object missiles generated; cars lifted off ground.
F3	158–206	*Severe damage*. Roofs and some walls torn off well-constructed houses; trains overturned; most trees in forest uprooted; heavy cars lifted off the ground and thrown.
F4	207–260	*Devastating damage*. Well-constructed houses leveled; structures with weak foundations blown away some distance; cars thrown and large missiles generated.
F5	261–318	*Incredible damage*. Strong frame houses leveled off foundations and swept away; automobile-sized missiles fly through the air in excess of 100 m (109 yards); trees debarked; incredible phenomena will occur.

reported before midnight. On May 4, tornadoes continued to develop, and over the next two days there were many more tornadoes, resulting in more than 140 over the three-day period. Tornadoes caused almost $2 billion in damage and killed 50 people (plate 11). More than 2,000 homes were destroyed and another 9,000 were damaged. Some small towns, such as Mulhall in central Oklahoma, were utterly destroyed, and Tanger Outlet Center in Stroud was obliterated and has never been rebuilt. The tornadoes even cut a wide swath of destruction through big cities such as Oklahoma City. Other tornadoes from this swarm struck Texas, Kansas, Arkansas, and Tennessee.

Of the 140 known tornadoes from this swarm, the Moore-Bridge Creek tornado was the biggest and most destructive. It struck around 7:30 p.m., on the evening of May 3. It was an F5 event on the Fujita scale of tornadoes (table 7.1), which ranks the largest possible storm, with winds at least 322 km/h

(200 mph). Winds were clocked at 484 km/h (381 mph), breaking the record for wind speeds ever measured in a tornado. This prompted a discussion about revising the Fujita scale to include an "F6" ranking. The tornado started near Amber, Oklahoma, and traveled in the usual northeasterly trend parallel to Interstate 44, striking Bridge Creek and reaching the southern suburbs of Oklahoma City. It then trended more northerly and hit more of Oklahoma City before striking Moore, Del City, and Tinker Air Force Base. It began to diminish near Midwest City and finally lifted off the ground a short distance to the north. When it was all over, 36 people had died, and more than 8,000 homes had been lost or damaged, causing more than $1.1 billion (in 2009 dollars) in destruction, making it the costliest single tornado in American history. It could have been one of the deadliest, but because of plenty of early warnings, most potential victims were safe in shelters. Those killed had no shelter or were in mobile homes that were picked up and thrown about. Three people died when they sought shelter under a highway overpass, which, despite the urban myths, does *not* provide adequate shelter (see the following discussion).

Kansas "Cyclones"

L. Frank Baum's 1900 book and the 1939 MGM movie *The Wizard of Oz* associate "cyclones" with Kansas. The term *cyclone* is an archaic word for tornado, but today, meteorologists apply it only to the tropical cyclones or hurricanes of the Indian Ocean. All spiraling storm clouds in the Northern Hemisphere, whether they are tornadoes or hurricanes, have a clockwise, or "cyclonic," sense of rotation when viewed from above. In the Southern Hemisphere, spiraling storm clouds move counterclockwise, or "anticyclonic," because of the difference in the Coriolis effect and the spin of the earth beneath fluid currents like water or air.

Baum was not exaggerating, because Kansas has had its share of "tornado alley" storms, second only to Oklahoma. Twenty years before he wrote *The Wizard of Oz*, Baum was apparently inspired by a truly immense tornado swarm that swept through Kansas on May 29–30, 1879. Newspaper headlines around the United States printed articles about "The Great Kansas Cyclone," which became popular usage and a well-known common memory for that generation. The worst tornado of that storm struck Irving, Kansas, on May 30, 1879, and destroyed the town, killing 66 and injuring 60 others (virtually the entire population). According to eyewitnesses, the storm picked up one house and whirled it around like a top (hence the similar events in Baum's story). The rest of the buildings were destroyed, leaving only foundations.

I have only experienced a few tornadoes in my years living in the region (mostly in South Dakota, Nebraska, and Illinois), but I have sought shelter many times during severe storm warnings. I am familiar with the sounds of the warning sirens that every town and city in the Great Plains and Midwest has set up for emergencies. I have never seen a tornado because I head for shelter when sirens sound. My wife's family is from the Wichita, Kansas, area and has experienced many storms. Several members of her family survived the May 25, 1955, tornado that wiped out the town of Udall, leaving only a few brick buildings standing (fig. 7.2). Luckily, they were in a farmhouse 4 km (2.5 miles) on the outskirts of town and were not killed. This F5 storm hit at night, killing 83 and injuring more than 270 people (out of only 610 residents), making it the deadliest tornado ever to hit Kansas. Every building of the 192 in the town was damaged and 170 of them were destroyed, including the water tower and the grain elevator. My father-in-law was 30 km (20 miles) away at the time, and even he could hear the thunderous sound of the tornado as it passed through Udall and made its way toward Derby, Kansas. Unknown to the Udall residents, just minutes earlier, the same tornado had devastated the town of Blackwell, Oklahoma, just south across the border, killing 20 and injuring 250.

Udall marshall Wayne Keely gave the following account:

The first warning we had was the noise. It was like one of those B1 bombers that are around these days coming into town. I got my family and the kids and got them into the cellar just before it hit. I remember my wife trying to light a candle, and myself not being able to breathe too well. I was able to get a quick glance out of the cellar while the tornado was over us. It looked like there was

Fig. 7.2. Damage from the Udall, Kansas, tornado that struck on May 25, 1955. A, Aerial view of the devastation. B, The frame of a 1952 Chevy pickup thrown into a tree. (Images taken by M. W. Keely)

electricity inside of it. I'm not sure if it was metal hitting together or maybe static electricity. Debris was coming into the cellar and blocked the entrance.

. . . There was a 1952 Chevy pickup in the tree in my front yard [fig. 7.2B]. The owner of the truck was found dead later outside of town. I walked through the damage. I remember not being able to tell what part of town I was in. We found another neighbor setting on a porch, setting down leaning up next to a pole like they were sleeping, but they were gone. (www.tornadochasers.com/udall)

This account was given by Boyd Binford:

As I was getting in the car my mom said is it a tornado? I was pulling my door shut at that time, the car was beginning to rock. I reached down to put the car in drive, as I did I made it to reverse, I think my mom must of had her foot on the gas at this time, because when the car went into reverse and started to take off and that's when it hit. The car begin to bounce up and down but was not moving. The car then spun around and was pushed into a ditch, the headlights were pointing directly at the house across the street. Then I saw the house blow away, then the car started going end over end. (www.tornadochasers.com/udall)

Western Kansas has also had its share of deadly storms. Greensburg, Kansas, has been struck multiple times, including devastating tornadoes in 1915, 1923, and 1928. One report comes from a farmer living outside the town who managed to look inside the funnel from the entrance of his shelter on June 22, 1928:

I looked up and to my astonishment I saw right up into the heart of the tornado!

There was a circular opening in the center of the funnel, about 50 or 100 feet in diameter, and extending straight upward for a distance of at least a mile, as best I could judge under the circumstances. The walls of this opening were of rotating clouds and the whole was made brilliantly visible by constant

Fig. 7.3. Damage to Greensburg, Kansas, after the May 4, 2007, storm. (Photo courtesy FEMA)

flashes of lightning which zigzagged from side to side. (*Monthly Weather Review*, 1930, 58:205)

Another tornado struck Greensburg, Kansas, on May 4, 2007, at 9:45 p.m. and was about 2.7 km (1.7 miles) wide and traveled for 35 km (22 miles). More than 95 percent of the city (fig. 7.3) was destroyed and the remainder was severely damaged. It was the first F5 storm since the May 1999 storms that hit Oklahoma, and the first to be rated EF5 on the Enhanced Fujita scale, which modified the original damage-based Fujita scale (table 7.1) accounting for wind speeds. Doppler radar clocked the wind speeds at 330 km/h (205 mph). Eleven people were killed and many more injured, but the death toll would have been much worse without the warning sirens that sounded 20 minutes before the storm arrived. Kansas Governor Kathleen Sebelius and President George W. Bush visited the scene and declared it a disaster area. Greensburg is now being rebuilt as a "green" city, following the LEED (Leadership in

Energy and Environmental Design) Green Building rating standards, which makes it the first green U.S. city. A nonprofit organization, Greensburg Green Town, has been set up to help the residents learn about and build in an environmentally sound way.

What Causes Tornadoes?

What turns an ordinary thunderstorm into a swirling twister? It all starts with the severe weather conditions that generate large cumulonimbus, or "thunderhead," clouds. Such thunderheads are usually produced when a mass of warm, moist air meets a mass of cold, dry air, generating a strong frontal system. If there is a strong high-altitude jet stream operating above it with winds of 150 mph or more, then the three moving air masses interact to form a strong wind shear that gets the thundercloud spinning.

With "tornado alley," the central plains have just the right combination of conditions to make frequent tornadoes (fig. 7.4). During the spring and summer, masses of cold, dry air come down from Canada, which meet warm, moist air from the Gulf of Mexico. At the same time, the jet stream shifts southward and focuses its energy over the midcontinent. Warm, dry air may come east from the desert Southwest. The Rocky Mountain front blocks the spread of these air masses to the west, so they end up concentrated in a narrow zone. Their meeting point is usually in the Great Plains, generating huge frontal systems week after week. If they are large enough, tornadoes can be expected. Tornadoes are most common during the afternoon and evening, when heat builds up during the day to reach a critical point and enough thunderstorms are produced. The boundary between the two frontal systems usually trends northeast to southwest, so tornadoes usually travel northeasterly as well.

Although tornadoes are known from every continent except Antarctica, they are far more common in the American Midwest than anywhere else. They have been documented in southern Africa, western Australia, New Zealand, south-central and eastern Asia, east-central South America, and northwestern and southeast Europe. In most instances, dynamics are similar: a broad

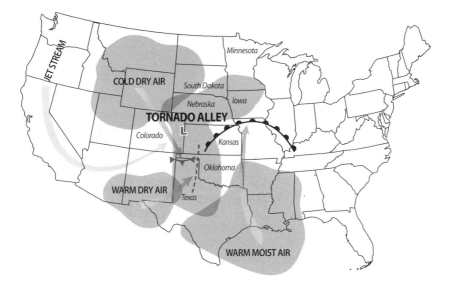

Fig. 7.4. Diagram showing the atmospheric dynamics of "tornado alley." (Diagram from NOAA)

plains region (often confined by mountains) where warm, moist air comes up from the tropics and meets cold, dry air coming down from polar regions.

Tornadoes are also common in Bangladesh, where they kill almost 200 people a year, the deadliest annual toll from tornadoes in the world. About 1,300 people were killed in the Daultipur-Salturia tornado in Bangladesh on April 26, 1989. Bangladesh has had more than 20 tornadoes that killed 100 or more people, more than half of the total known for the entire world. Like tornado alley in the American plains, the conditions in Bangladesh are ideal for tornado formation: cold, dry air coming down from central Asia meets warm, moist air from the Bay of Bengal, and the Himalayas confine the flow to the broad plains of the Ganges-Brahmaputra delta.

The tornado is actually the core of the spinning vortex within the cloud, which spins on a horizontal axis as one air mass slides beneath the other. A strong wind shear with updrafts of warm air and downdrafts of cold air tilts the horizontally rolling air into a vertical orientation. If the vortex extends down to the ground, it becomes visible as it picks up dust and debris (fig. 7.5). Most

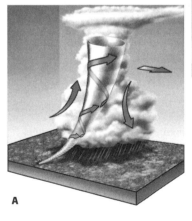

Fig. 7.5. A, Diagram showing the motion of air in a thunderhead that produces tornadoes. A warm cloud of moist air comes from the base and is overridden by cool, dry air from above. The two interact to form a swirling vortex, with warm air producing strong updrafts and cool air descending in strong downdrafts. These tilt the rotating cylinder of air from horizontal to vertical. The tornado we see is only the basal tip of a much larger spiraling vortex running through the entire cloud. (Diagram courtesy NASA; redrawn by Pat Linse) B, Image of a tornado developing from a cylindrical vortex spinning in a horizontal axis (*top*) and then shifting to a vertical orientation and eventually touching down. (Photo courtesy NOAA)

tornadoes have wind speeds of 64 km/h (40 mph) to 177 km/h (110 mph), and their funnels are approximately 75 m (250 feet) across. Some are much larger (more than 1.6 km, or 1 mile across) or much smaller (only a few meters) across. They may touch down and devastate only a few kilometers or miles, but some tornadoes have cut a swath of more than 500 km (hundreds of miles) in length. The name *tornado* may come from the Spanish word *tronada*, or "thunder-storm," possibly combined with the Spanish verb *tornar* (to turn) to give us the word "tornado."

Tornadoes cause damage in a number of ways. The most obvious is the high wind speed, which can lift almost anything, from houses and trucks and even larger objects, and carry or tumble them considerable distances. Large tornadoes can uproot and lift huge trees, rip asphalt off highways, and flatten buildings. Strong and violent winds frequently throw debris great distances, and flying debris made of broken lumber and metal can impale objects. Tor-nadoes have extremely low air pressure in their centers, which may cause windows to pop out of buildings or cause buildings to explode because air pressure inside is greater than air pressure within the tornado. Tornadoes can do weird and quirky things, including picking up cows and carrying them unharmed for hundreds of meters, plucking chickens of their feathers without killing them, driving thin pieces of straw through wood like nails, and picking up and carrying railroad cars off the ground and setting them down on their wheels again.

Tornado Myths

Tornadoes have been such terrifying forces of nature for so long that many myths and misconceptions have arisen about them. Some are harmless, but many are harmful if you make the wrong decision.

Where is the safest place to be during a tornado? As most midwesterners know, the best protection is a concrete-lined shelter with solid doors (like a bomb shelter) in the core of the building. Most midwestern houses have "tor-nado cellars," where the family can shelter, even if the house is ripped to pieces above them. If such a shelter is not possible, then a room in the lowest floor of the center of the house with some interior walls and few or no windows (such as a bathroom) is the best alternative. Contrary to the popular myth, the northeastern corner of the house is not the safest part of a structure, nor is the southwestern corner the most dangerous. This myth comes from an 1887 book on tornadoes by John Park Finley, and it was based on the observation that (at least in "tornado alley") most tornadoes travel southwest to northeast (because this is the trend of the weather fronts when the Gulf and Canadian air masses meet). Later research showed the opposite was true: there were fewer deaths in the southwestern "leading edge" part of a house than in the northeast "trailing edge." But neither is as safe as the center of the house.

Another myth is that highway overpasses provide adequate shelter. This was propagated when a TV film crew in El Dorado, Kansas, shot footage as they hid under an overpass during a 1991 tornado. However, their situation was unusual, because they were wedged in a shelflike crawl space at the top of the embankment with steel girders to hang on to, and the tornado did not actually pass over them. Nevertheless, the sensational footage was broadcast all over the country and led many people to believe they were safe under an overpass. Unfortunately, this myth proved deadly because several victims of the 1999 Oklahoma tornado storms tried to shelter beneath overpasses and were killed by flying debris or swept into the tornado. An overpass may actually be one of the *worst* places to take shelter during a tornado because overpasses are elevated aboveground, and wind speed is faster on higher terrain. The confined space may create a wind-tunnel effect, focusing the force of the winds into a narrow channel, strengthening them. If you stop your car in an underpass, you might block traffic and make the situation worse for others whose cars are left in exposed positions.

Some people think driving as fast as possible to outrun a tornado is safer than staying in a shelter. The opposite is true. Your car may be able to outrun a tornado if you drive in a straight unimpeded line, but if the tornado changes course and blocks the roads that provide the only escape, you may find yourself cut off or trapped. Tornadoes drop debris and fell power lines and telephone poles, which can block roads. Tornadoes are usually accompanied by intense rain, hail, and flying dust storms, making driving slippery and visibility limited, so you are more likely to be in an accident. Worst, drivers attempting to escape a tornado may be panicky and careless and more likely to crash or cause a traffic jam, making other drivers vulnerable. In Wichita Falls, Texas, on April 10, 1979, a number of people trying to flee a tornado got caught in traffic and were killed in their cars. If you are in a car and see a tornado bearing down, the National Weather Service recommends you seek shelter in a ditch or culvert, but do not sit in a stalled car.

A common belief is that tornadoes are "attracted" to trailer parks. Tornadoes are not affected by any human structures on the ground. They move in a more or less steady path, depending on the winds that produce them. A higher percentage of deaths (about 50%) during tornadoes occurred to victims in trailers, but that's not because tornadoes are attracted to them. Trailers and mobile homes are simply more vulnerable because they are light, airtight, and not anchored to the ground on a solid foundation; therefore, a strong tornado can easily lift them. High winds can lift, demolish, or roll trailers and mobile homes in ways that other structures cannot. In many parts of the world where land is cheap and permanent housing expensive, a neighborhood may have many trailers, which are vulnerable when tornadoes strike.

Another myth is that tornadoes do not strike big cities. Several major cities, including Oklahoma City and Salt Lake City, have been hit by tornadoes. From spring 1997 to spring 2000, approximately 3,600 tornadoes occurred in the United States, and 10 of them struck large metropolitan areas: Miami, Florida; Birmingham, Alabama; Nashville, Tennessee; Little Rock, Arkansas; Cincinnati, Ohio; Wichita, Kansas; Salt Lake City, Utah; Milwaukee, Wisconsin; and Fort Worth, Texas, were among them. Oklahoma City has been hit more than 109 times between 1893 and 1999. Many larger cities produce "heat islands" from warm plumes of exhaust and concrete absorbing and reflecting heat, which may help to deflect some thunderstorms. But there is a much simpler explanation for why tornadoes rarely hit big cities. There are few large cities in the Great Plains and the Midwest, and these cities do not occupy a large area of land. They are simply small targets for a typical random array of tornadoes to strike.

The Super Outbreak, 1974

Although tornado alley states like Texas, Kansas, and Oklahoma have the lion's share of annual tornadoes in the United States, the largest single-day outbreak of tornadoes did not occur in these states but from Alabama and Georgia up to Michigan and Ontario (fig. 7.6). In one day (April 3, 1974), 148 tornadoes occurred in 13 states, damaging 1,440 km² (900 square miles) of land. There were 6 F5 and 24 F4 tornadoes, an all-time record, and at one point, there were 15 active tornadoes on the ground simultaneously. The situation was so chaotic and the multiple storms so hard to track that officials issued a blanket tornado warning for the entire state of Indiana. The individual paths of those 148 tornadoes formed a combined path length of 4,160 km (2,600 miles). This single deadly day killed 335 people and hospitalized 1,200 more. More than 7,500 houses were destroyed and 6,000 more were damaged, including another 2,100 mobile homes destroyed and more than 4,000 farm buildings and 1,500 small businesses. More than 27,600 families suffered significant losses, making it one of the costliest disasters in American history, totaling about $3.5 billion (in 2005 dollars).

On April 2, a series of weather conditions brewed into the perfect storm. A big cold front spread east from the Rockies toward a low-pressure system over Michigan, which also sucked in humid air from the warm Gulf of Mexico. A strong polar jet stream flowed across the Midwest, with a big portion flowing right over the stretch from Texas to New England. Finally, warm, dry desert air from the Southwest moved up the Mississippi and created an inversion layer, with warmer air on top of cooler air. Since heat rises, this inversion condition is stable and stagnant but also prone to bad side effects when other weather conditions prevail. When these weather systems collided on April 3, the Gulf air was forced up through the inversion layer, producing huge anvil-shaped thunderheads, which were then sent spinning by their own motion and the

Fig. 7.6. One of the 148 "super outbreak" tornadoes of April 3, 1974, as it struck Xenia, Ohio. (Photo courtesy NOAA)

strong jet stream wind shear. By 1 p.m., the greatest barrage of tornadoes ever recorded had begun.

Of the many different tornadoes and many different towns involved, a few instances stand out. The little town of Xenia, Ohio, was hit by an F5 storm formed when two smaller tornadoes merged to form one giant funnel cloud (fig. 7.6). It flattened nearly every building in Xenia, picked up railway cars, and toppled gravestones. It killed 32 people and injured about 1,150 others, causing $150 million in damage. President Nixon visited afterward and declared it a Federal Disaster Area. Before the storm, there were no warning sirens, but now there are 20—too late to have done any good in the 1974 storm. Another F5 tornado hit Brandenburg, Kentucky, and destroyed nearly every building, while killing 31 people (18 on a single block). Once again, there were no warning sirens in place, and the only warning came when a local disc jockey looked out the window of the studio and saw the tornado coming. He told listeners to take cover, and then the radio station was obliterated. About an hour later, the same thunderhead produced an F4 tornado that hit Louisville, Kentucky, where it demolished much of the city on a 35 km (22 mile) long swath. Luckily, the local weather station and its traffic helicopter were able to follow and track the storm and warn listeners, so there were only a few deaths in the Louisville area.

Across the Ohio River, DePauw and Madison, Indiana, were hit by two separate F5 tornadoes, while Monticello, Indiana, was hit by an F4 event, and Cincinnati, Ohio, was hit by a different F5 event. Numerous towns in the South were struck by different tornadoes: Huntsville, Guin, Jasper, and Tanner, Alabama. Finally, the day's misery became international when an F3 tornado struck Windsor, Ontario (across the border from Detroit, Michigan). This storm killed 9 and injured 20, almost all trapped inside a curling rink.

Attack of the Killer Tornado, 1925

The 1974 "super outbreak" may have been the largest cluster of storms in a single day, but the prize for the biggest and deadliest tornado ever goes to the infamous Tri-State Tornado of 1925 (fig. 7.7A). It did not hit the usual tornado targets in Texas and Oklahoma but instead cut a wide swath across Missouri, Illinois, and Indiana. After 1 p.m. on March 18, 1925, witnesses reported a sound "like a thousand freight trains" headed their way. It was one of the largest and broadest tornadoes ever seen, with a path several kilometers (several miles) wide, moving northeast across Missouri at about 100 km/h (60 mph),

Fig. 7.7. A, Map of the course of the Tri-State tornado, 1925. (From Wilson and Chagnon, 1971. Courtesy Illinois State Water Survey; redrawn by Pat Linse.) B, Ruins of the Longfellow School, Murphysboro, Illinois, where 17 children were killed. The storm hit the school at about 2:30 p.m., just as the school day was winding down. C, Devastation of the town of Griffin, Indiana, during the 1925 Tri-State tornado. (Courtesy Wikimedia Commons)

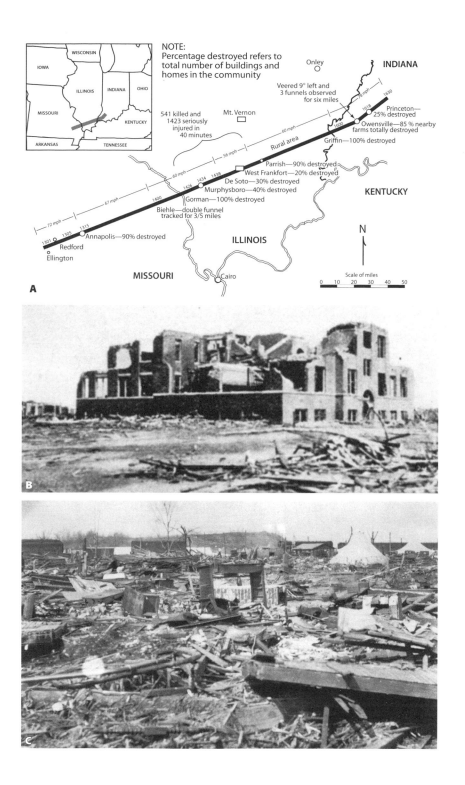

NOTE:
Percentage destroyed refers to total number of buildings and homes in the community

Onley

INDIANA

Veered 9° left and 3 funnels observed for six miles

78 mph

1630

1618

Princeton—25% destroyed

1600

Owensville—85% nearby farms totally destroyed

541 killed and 1423 seriously injured in 40 minutes

Mt. Vernon

60 mph

Rural area

Griffin—100% destroyed

KENTUCKY

56 mph

Parrish—90% destroyed
West Frankfort—20% destroyed
De Soto—30% destroyed
Murphysboro—40% destroyed
Gorman—100% destroyed
Biehle—double funnel tracked for 3/5 miles

60 mph

1434 1438
1426

1400

67 mph

72 mph

1315

Annapolis—90% destroyed

ILLINOIS

N

1301 1305
Redford
Ellington

MISSOURI

Cairo

Scale of miles
0 10 20 30 40 50

A

B

C

faster than any car or horse could outrun. Witnesses saw the huge formless round cloud of dust moving toward them and did not recognize it as a tornado. It was difficult to judge its speed, so many people did not realize how fast it was moving and were overcome before they could escape.

After leveling Annapolis, Missouri, and many other smaller towns along the way (killing 11), the tornado crossed the Mississippi River into Illinois, where it struck the town of Murphysboro (fig. 7.7B). Every building in its path was destroyed, and the rest of the town caught fire from the open flames left behind. The broken water mains left firefighters watching helplessly as the town burned down. Two hundred and thirty-four people died. It hit De Soto, Illinois, where the residents watched the terrifying huge cloud carrying houses, cars, and uprooted trees racing toward them. It hit the town school, lifting off the roof and knocking down the walls. Of 125 students and teachers, 88 died. The dead students were laid out on the lawn afterward, but their parents could not claim them, because they were dead, too. A train engineer was on the tracks headed straight toward the funnel cloud. He decided to speed up and try to race through it, which he did—except that the roofs of the train cars were stripped off.

In 40 minutes, the tornado had struck dozens of towns in southern Illinois, killed 541 people, and injured 1,423. Finally, the killer crossed the Wabash River into Indiana, where it leveled Griffin (fig. 7.7C), Owensville, and Princeton, killing another 71 people.

More than 700 died, making it the deadliest single tornado in U.S. history, and more than 2,000 were injured. It hit nine schools in three different states and killed 69 students, the highest death toll of schoolchildren in any natural disaster (fig. 7.7B). Four towns were obliterated and never rebuilt. This catastrophe caused at least $1.4 billion (in 2007 dollars) in damage and severe effects on the local economy. Many starving and homeless people were left behind. Anarchy, crime, and looting were widespread. Recovery was slow, and the entire region did not recover economically or socially for many years after the event.

FOR FURTHER READING

Akin, W. E. 2002. *The Forgotten Storm: The Great Tri-state Tornado of 1925*. Lyons Press, Guilford, CT.

Ball, J. A. 2005. *Tornado! The 1974 Super Outbreak*. Bearport Publications, New York.

Bechtel, S., and T. Samaras. 2009. *Tornado Hunter: Getting Inside the Most Violent Storms on Earth*. National Geographic, Washington, DC.

Bluestein, H. B. 1999. *Tornado Alley: Monster Storms of the Great Plains*. Oxford University Press, New York.

Felknor, P. S. 2004. *The Tri-state Tornado: The Story of America's Greatest Tornado Disaster*. I-Universe, New York.

Grazulis, T. P., and D. Flores. 2003. *The Tornado: Nature's Ultimate Windstorm*. University of Oklahoma Press, Norman.

Hollingshead, M., and E. Nguyen. 2008. *Adventures in Tornado Alley: The Storm Chasers*. Thames & Hudson, London.

Laffoon, P. 1975. *Tornado*. Harper & Row, New York.

Levine, M. 2007. *F5: Devastation, Survival, and the Most Violent Tornado Outbreak of the Twentieth Century*. Miramax, New York.

Mathis, N. 2007. *Storm Warning: The Story of a Killer Tornado*. Touchstone, New York.

Weems, J. E. 1977. *The Tornado*. Doubleday, Garden City, NY.

Whipple, A. B. C. 1982. *Storm*. Time-Life Books, Chicago.

Blizzards

White Death

A blinding barrage of snowflakes was driven by strong winds which wrapped
schoolgirls' skirts around their legs, frightfully impairing efforts of the
youngsters to reach safety. Battling the powerful force of the stiff cold air
currents, and stumbling through the reduced visibility of the fierce snowy gale,
young Avis "was down more than up," and her hands froze painfully
during the lengthy one-block trip to her house.
—Dick Taylor, *The Schoolhouse Blizzard*

The "Great Blizzards" of the 1880s

During the 1870s, settlers flocked to the Great Plains and Rocky Mountain
regions by the thousands. The completion of the transcontinental railroad in
1869 and numerous other railroads opened access to homesteaders and farmers. Meanwhile, wars with indigenous Americans had been winding down
ever since the 1876 Battle of Little Bighorn spurred the U.S. Army to force
Native Americans onto reservations. Hunters had been steadily slaughtering
the bison herds at the rate of millions a year, so that by the early 1880s, the
great sea of bison that once covered the plains were nearly extinct. Civiliza-

tion was rapidly coming to the Wild West, and with the invention of barbed wire, even the long cattle drives to railheads like Dodge City, Kansas, were coming to an end.

The railroad companies had made huge profits off the land Congress had originally given them. Along with many other land hucksters and swindlers, they heavily advertised the Great Plains and Rockies as a paradise ready to be farmed, plowed, and turned into fertile green valleys. These misleading promotions claimed that the climate was mild and the lands well watered, even though as early as the 1860s, John Wesley Powell had argued that "The Great American Desert" would never be wet enough to support full-scale agriculture without dams and irrigation. The popular myth was that "rain follows the plow," so if a farmer plowed the land, it somehow released moisture and triggered more frequent rain in formerly dry country. In the words of promoter Charles Dana Wilber:

God speed the plow . . . By this wonderful provision, which is only man's mastery over nature, the clouds are dispensing copious rains . . . [the plow] is the instrument which separates civilization from savagery; and converts a desert into a farm or garden . . . To be more concise, Rain follows the plow. (The West Film Project and WETA 2001)

During the 1870s, an abnormally long wet spell made the Great Plains and Rockies unusually green, and farmers and cattlemen were able to sustain themselves. It turned out that this wetter climate was short term, and by the early 1880s, the West began to revert to its normal hot, dry summer conditions. Soon, cattle were starving all over the plains and dying from lack of fodder. Ranchers and farmers who had homesteaded the region saw their investments wiped out.

The final blow came when the climate reverted not only to its normal dry conditions but also went through a cold snap in the 1880s that produced blizzards and freezing conditions that lasted months. This was the last gasp of the Little Ice Age, a climatic cooling cycle that had chilled much of the Northern

Hemisphere since the 1600s. The winter of 1880–1881 was particularly harsh, with blizzard after blizzard killing hundreds and stranding settlers who suffered extreme privation and starvation. These events were vividly portrayed by Laura Ingalls Wilder, author of *Little House on the Prairie* and seven other books in the Little House series. In *The Long Winter*, Wilder described how her snowbound family nearly froze and starved to death while stranded in De Smet, South Dakota, for many months. The blizzards began in early October before most of the crops had been harvested and never seemed to let up, lasting for two or three days each with only a few days of clear, cold weather in-between. When the trapped trains finally were able to break through in May, the residents of De Smet celebrated Christmas (and their survival)—and their Thanksgiving turkey was still frozen. The terrible winter of 1880–1881 is also described in O. E. Rolvaag's *Giants in the Earth*, an award-winning tale of immigrant Norwegian farmers in the Dakota Territory, who suffered horribly during the blizzards.

The final blow came during the winters of 1886–1888. The previous droughts and hard winters had put the cattle business in jeopardy, and the herds were thin and weak. Overgrazing had depleted the once fertile grasslands of the Great Plains, and sheep were competing with the cattle for grass. Beef prices were falling, so ranchers were on the edge of financial disaster. The first snow came on November 13, 1886, and did not let up for a month. Then, in January 1887, the Great Plains were pounded by incredible blizzards, with howling gale force winds and temperatures well below freezing. The cattle, which before had always been able to survive in the open range during winter, were trapped by the deep snow and unable to find shelter to survive the days of subfreezing winds and blizzards (figs. 8.1, 8.2). Some cattle were too weak to stand and were literally blown over. Others froze to death as they stood, buried deep in snow. The rancher Teddy Blue Abbott wrote, "It was all so slow, plunging after them through the deep snow that way . . . The horses' feet were cut and bleeding from the heavy crust, and the cattle had the hair and hide wore off their legs to the knees and hocks. It was surely hell to see big four-year-old steers just able to stagger along" (The West Film Project and WETA 2001).

Fig. 8.1. During the blizzards of the 1880s, this scene of a dying steer freezing in a blizzard was a common sight. (Courtesy NOAA Photo Library)

Finally, the spring thaws came, and with them a grisly scene of devastation. Hundreds of carcasses of cattle were spread across the range, slowly thawing and then rotting in the sun, as wolves, coyotes, and vultures feasted. As the snowmelt continued, rivers flooded, and the bloated carcasses of cattle clogged the rivers as they floated downstream. Rancher Lincoln Lang wrote, "[I saw] countless carcasses of cattle going down with the ice, rolling over and over as they went, sometimes with all four stiffened legs pointed skyward. For days on end . . . went Death's cattle roundup" (The West Film Project and WETA 2001). The winters of 1886–1887 and 1887–1888 came to be known by the ranchers as "The Great Die-Up" because so many livestock lost their lives. Most of the ranchers were wiped out and sold their land claims and their remaining herds to recoup their losses and find another way to make a living.

With these events, the great age of big, open-range cattle ranches on the plains was transformed, as surviving ranchers learned to cope with smaller

Fig. 8.2. The great Dakota blizzard of March 1966 was almost a repeat of the 1886–1888 blizzards. Shown here are hundreds of cattle dead from freezing, Brookings, South Dakota. (Courtesy NOAA Photo Library)

herds and smaller confined range lands, or diversified into other businesses. The era of open-range cattle drives and the corresponding rancher lifestyle, which future president Theodore Roosevelt once romanticized as "the pleasantest, healthiest and most exciting phase of American existence" (Roosevelt 1888, 24) was no more. Roosevelt had come to the North Dakota Badlands to make a living as a rancher in 1884. He fled there to heal from his grief and bereavement after his wife and mother had died on the same night, February 14, 1884. Roosevelt became a successful and colorful rancher for three years, but events soon drove him out of the West. Like nearly everyone else, Roosevelt lost his $60,000 investment in his ranch during the winter of 1886–1887, sold out, and returned East. Upon his return, he remarried and resumed his political career, but his ranching days and "Roughrider" friends were always part of his past and his image.

One of the final events of the terrible plains winter of 1887–1888 was the Schoolhouse Blizzard, or Children's Blizzard. After a snowstorm across the Great Plains on January 5 and 6, 1888, temperatures warmed up remarkably to above freezing on January 12, 1888, as warm moist Gulf air moved over the northern plains and upper Midwest. Then a huge Arctic cold front swept down from the north, and temperatures dropped to –40°C (–40°F) in a matter of hours. It struck during the morning of January 12 in Montana, by midday in the Dakotas, and in the early afternoon in Nebraska. Children were still in school, so the sudden blizzard trapped many of them in their one-room country schoolhouses. Because it had been so mild (for January) and the storm had hit suddenly, many people were trapped far from their homes without adequate winter clothing. More than 500 people died of hypothermia, trapped in the snow and unable to reach shelter. The storm lingered for days, shutting off train travel and other transportation, so it was some time before all the frozen corpses were found and retrieved.

There are many accounts of the Schoolhouse Blizzard. One is from the historian of Pawnee County, Nebraska, Dick Taylor:

> On her way home from elementary school classes on Thursday afternoon, January 12, at Table Rock, Neb., 11-year-old Avis Dopp was caught in the fury of the violent winter assault . . . Flora Dopp, a former nurse, [later] revived her suffering daughter's agonized hands in cold water, while the girl believed her fingernails were starting to come off.
>
> . . . The winter sky cleared just before dusk, Avis later remembered, after three or four feet of snow had fallen and the strong winds had caused drifting. In the aftermath of the massive blizzard, Seymour H. Dopp kept his 17 pupils protected overnight in the country schoolhouse, where they had stockpiled fuel for a warm fire, and sheltered his faithful riding horse in the chilly enclosure of an unheated cob shed. (www.wintercenter.homestead.com)

The stories of the Schoolhouse Blizzard were tragic. In Plainview, Nebraska, the schoolhouse ran out of fuel for heating and began to freeze about 3 p.m. Teacher Lois Royce attempted to take her students to her own boardinghouse, only 75 m (82 yards) away. But the snow and wind were so bad and the visibility so poor that they all became lost over that short distance and froze to death. The teacher survived, but her legs were frostbitten and had to be amputated. In Holt County, Nebraska, Etta Shattuck sought shelter in a haystack where she remained for three days. When she was eventually rescued, she was still alive, but later died from complications of the surgery to amputate her frostbitten limbs. In Mira Valley, Nebraska, teacher Minnie Freeman safely led her 13 students through the blizzard to her home 0.8 km (0.5 miles) away, possibly by tying them together with a clothesline so they would not get lost or separated. Many other children across the Great Plains were not so lucky, since about half the total casualties (235 people) were children who died in their schools or trying to get home from them.

The storms of that winter were not just restricted to the plains. The eastern part of the country suffered the most extreme snows in American history in March 1888, an event known as the Great Blizzard of 1888. Early March had already started with mild spring weather, so many Americans thought that winter was finally over. Shortly after midnight on March 12, 1888, the heavy rains of the evening turned to snow as the temperatures plummeted, and soon snow was falling all over the Mid-Atlantic states and New England (from Virginia to Maine) and in Maritime Canada at an incredible rate. By the next day, the snow was piled up at least a meter high nearly everywhere, with the maximum snowfall of 1.4 m (58 inches) dropped in a single storm. Winds reached up to 130 km/h (80 mph), blowing the snow into huge drifts and producing whiteout conditions all over the Northeast. The snowdrifts soon piled up to 15 m (50 feet), burying cars, trains, and buildings under thick snow.

When the Great White Hurricane finally ended on March 14, 1888, the entire Northeast coast was paralyzed (fig. 8.3). All telegraph and telephone lines were down, isolating cities and preventing the communication of important emergency information. In future years, cities began to put their wires

Fig. 8.3. Images of the snowdrifts after the Great Blizzard of 1888. A, New York &
Harlem Railroad train trapped in snowdrifts near Coleman Station, March 13, 1888.
B, The tracks and trains at 45th Street and Grand Central Station were paralyzed.
(Images courtesy NOAA Photo Library)

underground to prevent this problem from recurring. All the railroads and streetcars were also gridlocked, and the roads were clogged with snow, preventing anyone from traveling for days. The snowdrifts across the trains up from New York to Connecticut took eight days to clear, which led to the development of the Boston and New York City subway systems. Fire stations were also locked in snow, so there was no response to fires, and more than $25 million in damage was caused by fire. As the weather rapidly warmed in late March, the snow melted and flooded many regions, including Brooklyn. There was so much snow on the ground that attempts were made to load it up and dump it in the ocean. At least 400 people died from the snow and its side effects, such as exposure and frostbite, including 200 just in New York City.

The severe winters of the 1880s were never quite duplicated in subsequent American history, because the overall trend in climate since then has been dominated by global warming. These storms were the final gasp of the Little Ice Age before it came to an end. But their effects were profound. Not only did these blizzards kill many people and cattle, but they also changed the way of life in the American West and ended the practice of large-scale open-range cattle ranching. They slowed the flood of settlers and homesteaders in the northern plains, since the myth of the "mild weather" of the region was shown to be a promoter's lie. The westward migration did not halt, of course, but it shifted to other areas, especially Oklahoma, which had been reserved for Native Americans but had been opened in the 1880s and 1890s to settlers who didn't want to experience the harsh winters of the northern plains.

What Causes Blizzards?

Blizzards are more than heavy snowstorms. They form when a high-pressure ridge of cold air from higher latitudes sweeps down and meets a low-pressure cell of moist air, usually from the subtropics. If the air is warm enough, these conditions produce thunderstorms and tornadoes, as discussed in chapter 7. During the winter, air is much colder, and precipitation comes down as snow, ice, or hail. A blizzard must have winds of at least 60 km/h (37 mph), temperatures well below freezing ($-7°C$ or $20°F$ or lower), and lots of falling or blowing snow. Typically, falling snow creates near whiteout conditions, cars are stranded, and people are subject to frostbite or hypothermia and death. Blizzards can cause unusually high death tolls, especially if the blizzard arrives suddenly and traps inadequately clothed people in exposed areas where they cannot stay warm.

In some parts of the world, extreme temperature change is not uncommon. I have been in the northern High Plains on a hot humid spring or fall day in which the outside temperatures were in the high 30 degrees Celsius (90 degrees Fahrenheit), mainly due to warm moist air from the Gulf. Then an Arctic air mass moves quickly over the plains, and the temperature drops below freezing in a matter of hours. If you are caught out in the field a long way from vehicles or shelter, wearing only your light summer clothing, you risk freezing to death in a few short hours if you cannot find warmth and shelter. This extreme change in temperatures can only happen in the continental interiors because these regions have no moderating effects of large bodies of water to absorb or release heat to balance the temperature change. By contrast, regions near the ocean have relatively mild stable weather because of the high heat capacity and thermal stability of water.

Blizzards can be costly disasters. Blinding snowfall causes traffic accidents due to reduced visibility and slippery roads, and chain-reaction collisions can occur during heavy snowstorms. People trapped in their stranded cars are at the mercy of the elements unless they can be rescued. When I lived in the northern United States, I carried the essential equipment for winter survival in my car: windshield scraper, snow chains, and a survival kit with water, blankets, food, a small chemical heater, and first-aid supplies. During a severe blizzard, traffic may be stalled and airports may close. Businesses and schools shut down for a "snow day" because it's difficult to conduct normal routines. These disruptions often cost millions of dollars in lost worker productivity and product delivery delays. In addition, the cost of snow removal can be considerable in areas of the country with heavy annual snowfall.

Even more striking is what happens when unusually cold snowy weather strikes the South or other regions where snow is rare. Traffic accidents increase

because drivers have little or no experience with snow driving. Many munici-
palities lack snow removal equipment. Frostbitten northerners or midwesterners may watch the Weather Channel or CNN with amusement, viewing
reports of Georgia or Alabama residents spinning and swerving in the snow,
but these images are not funny because these regions and their residents face
great danger due to their inexperience with snow.

The "Storm of the Century," 1993

Although the great blizzards of the 1880s were among the largest and deadliest in American history, major snowstorms happen every winter in northern
North America. Each year brings a few more blizzards. However, some winter
storms are more eventful than others, such as the White Hurricane of 1993,
one of the biggest and deadliest snowstorms, also known as the '93 Super-
storm, or the Storm of the Century. It struck a large area from Central America to eastern Canada, but the East Coast of the United States was hit most
severely. It was unique in historic storms for its intensity, size, and far-reaching
effects. The Deep South received record amounts of snow, with 30 cm (12
inches) falling in Alabama and 41 cm (16 inches) falling elsewhere in the
South. The northern parts of Florida were covered with 10 cm (4 inches) of
snow for the first time in recorded history. Hurricane force winds that accompanied this freakish snowfall sometimes reached 160 km/h (100 mph) in
places. Low-lying coastal regions in Louisiana and in Cuba were clobbered by
storm surge waves that washed many people away. Scattered tornadoes were
reported in many regions.

It all started with an unusually intense low-pressure system centered over
Michigan and another over Florida, which pulled a mass of warm, moist air
from the Gulf, with a line of thunderstorms along its leading edge. A huge,
cold air mass flowed down from the Arctic to the eastern United States, where
they both collided in one massive front (plate 12). Driving all this along was
an unusually strong jet stream that deflected southeastward over the southern
plains to create a strong cyclonic circulation pattern. By 1993, the National
Weather Service was sophisticated, and five days in advance of the storm, the

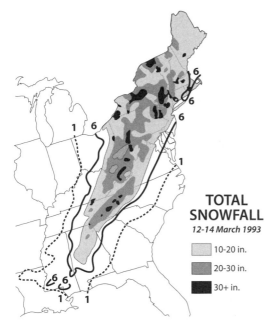

Fig. 8.4. Snowfall map from
the "Storm of the Century,"
March 12–14, 1993. (Map
courtesy NOAA; redrawn by Pat
Linse)

**TOTAL
SNOWFALL**
12-14 March 1993

10-20 in.

20-30 in.

30+ in.

service began issuing warnings of extremely heavy snowfall in the plains. In
the northeastern states, a state of emergency was declared before the snow
began to fall, freeing up resources and manpower, and keeping people indoors. In southeastern United States, the early March temperatures were
balmy, and local authorities did not believe the forecasts, so they ignored them
and did not prepare, nor warn their citizens. The weather forecasters proved
to be right after all.

By the evening of Friday, March 12, the temperatures had dropped rapidly
over the entire eastern United States, and the intense rainfall from the frontal
clouds quickly turned to snow. By Saturday morning, it was snowing heavily
from Florida to the southeastern part of the United States. Through the next
several days, the center of the low-pressure cell and the locus of heaviest snow
shifted northward until it hit New York, New England, and Maritime Canada
by Monday morning (fig. 8.4). Record low temperatures were recorded in Florida, Georgia, and the Carolinas, and even in the Northeast, the temperatures
were unusually low for mid-March, as low as –10°C (14°F) in New England.

The strong front spawned 50 tornadoes in Florida and Cuba, which killed more people than Hurricane Andrew in 1992, in which 45 people died. The storm surge wave that swept ashore in Florida and in the Caribbean surprised many and killed a few who were not expecting such large waves.

East Texas had snowy thunderstorms, and huge amounts of snow dropped up and down the eastern seaboard, with more than a meter (more than 3 feet) in many places, and snowdrifts as deep as 11 m (36 feet). The total volume of snowfall was calculated at 54 km³ (13 cubic miles), about 27 billion tons. Observers reported the lowest barometric pressures they had ever recorded, with values that typically accompanied the most intense hurricanes.

The severe weather caused at least 300 deaths, and more than 10 million people lost electricity for days. About 130 million people experienced this storm, or nearly half of America's population at the time. The Deep South was shut down for days because state governments had no emergency plans to handle an extreme storm and no snow removal equipment (plate 13). Chattanooga, Tennessee, was covered with a record 1.2 m (4 feet) of snow, and Birmingham, Alabama, received a record-shattering 43 cm (17 inches). Nearly every city in the South was paralyzed and shut down. Most residents expected the start of spring, so these events had a deep psychological impact. NASCAR races were canceled or rescheduled for a few weeks following the snowstorm. Weather conditions throughout the South were unsuitable for racing and for large crowds to travel to the races. In the South, roofs not built for the weight of heavy, damp snow collapsed under the snowfall. Buildings lost walls to the heavy winds and cantilevered decks insufficiently attached to the wall (plate 13). Hundreds of people hiking the Appalachian Trail had to be rescued because they lacked snow equipment, and trail shelters were inadequate for 24 hours of below-freezing temperatures, high winds, and 1 m (39 inches) of snow on the trails and mountainsides.

Farther north, snowy Syracuse, New York (which gets many meters of winter snow due to the lake effect), got a record 1.1 m (43 inches) of snow. New York City, which had seldom experienced March winter snow, received 12 cm (1 foot) of the white stuff. Heavier amounts, ranging up to 60 cm (2 feet) of snow, hit the suburban areas of Connecticut, New Jersey, and Pennsylvania, all records for this late in March. The snow was just as heavy in Maritime Canada, as were the storm surges, which clobbered the fishing fleet and killed 48 sailors. The storm caused about $6.6 billion in damage, making it one of the most costly blizzards in American history. New Englanders may point to the blizzard of 1978 as more severe in their region, while the blizzard of 1996 was more severe in the mid-Atlantic states; however, for sheer size, volume, and destructiveness, the 1993 snowstorm was truly the Storm of the Century in North America.

FOR FURTHER READING

Haraden, C. J. 2003. *Storm of the Century*. Times Square Books, New York.
Laskin, D. 2005. *The Children's Blizzard*. Harper Perennial, New York.
Tougias, M. 2002. *The Blizzard of '78*. On Cape Publications, Boston.

Ice Ages

Frozen Planet

Some say the world will end in fire,
Some say in ice.
From what I've tasted of desire
I hold with those who favor fire.
But if it had to perish twice,
I think I know enough of hate
To say that for destruction ice
Is also great
And would suffice.
  —Robert Frost, *Fire and Ice*, 1920

Outrage in Neuchâtel

In 1837, the Schweizerische Naturforschende Gesellschaft (Swiss Society of Natural Sciences) held its annual meeting in Neuchâtel, Switzerland. Surrounding the town were the towering cliffs of the tightly folded rocks of the Jura Mountains (from which the "Jurassic" Period got its name) and deep U-shaped valleys where remnants of glaciers could still be seen (fig. 9.1).

Fig. 9.1. Zermatt Glacier in the Swiss Alps, illustrated by Louis Agassiz in his 1840 book on the ice ages. Note the people for scale. (From Agassiz 1840)

Already one of the foremost scientific societies in Europe, the Schweizerische Naturforschende Gesellschaft was founded in 1815 to bring together the many Naturforschende Gesellschaften that had sprung up since 1746 in Zurich, Bern, Geneva, Basel, and most other larger Swiss cities. These societies maintained their own libraries and collections, and some even had botanical gardens. Their meetings and publications spread the development of natural history in Europe. The most prominent Swiss thinkers as well as scholars of many other nationalities participated.

The Neuchâtel meeting was like most others, with many presentations and reports on discoveries in botany, zoology, geology, mathematics, physics, and chemistry. A pioneering paper by Amanz Gressly on the important geological concept of sedimentary facies was presented but was soon overshadowed by the presentation of Louis Agassiz, a 36-year-old paleontologist and professor at the University of Neuchâtel (fig. 9.2). He was expected to give yet another talk on his continuing research into fossil fishes. Instead, Agassiz shocked his audience with a bold claim that the great glaciers still visible in the Swiss Alps had once covered all of Europe during an ice age, or *Eiszeit*. Distinguished scholars at the meeting were appalled and scandalized because the prevailing dogma maintained that the mysterious "drift" deposits were left by Noah's Flood, not by glaciers.

Agassiz and his critics quickly organized a postmeeting field trip to the Alpine glaciers to see the evidence for themselves. (This could never happen today in the strictly run and structured scientific meetings, where unbreakable plane reservations prevent anyone from planning a spontaneous postmeeting field trip.) Some prominent senior geologists in the world, including Elie de Beaumont and Leopold von Buch, viewed the evidence Agassiz had described, but they remained skeptical. Agassiz, in turn, was publicizing the thoughts of many other alpine experts, such as Ignaz Venetz, Karl Friedrich Schimper, and Jean de Charpentier. These men had worked on the idea of an ice age for years and had shown Agassiz how glaciers had once spread across Europe and left huge out-of-place boulders called *erratics* all over the landscape.

Fig. 9.2. Portrait of Louis Agassiz near the Unteraar Glacier, painted by Alfred Berthoud. (Courtesy Wikimedia Commons)

Agassiz followed his 1837 paper with an 1840 book, *Etudes sur les glaciers* (*Studies on Glaciers*). He not only outlined the evidence for how alpine glaciers moved and what deposits they left behind but went beyond his colleagues and mentors who thought that only individual lobes of glaciers were needed to explain the erratic boulders far from their source area. Agassiz argued that thick ice sheets had once covered all of Europe. That same year he visited Great Britain and went on geological excursions with the prominent geologists William Buckland and Charles Lyell. He accompanied Buckland through Scotland, where they saw many examples of glacial deposits. Soon, both Buckland and Lyell were converts. Most European scientists were unconvinced. They had no direct experience even with Swiss alpine glaciers, let alone with continental-scale glaciers, and could not imagine their neighborhood under a mile-thick pile of ice. In 1852, the explorer Elisha Kent Kane returned from a harrowing voyage with a thrilling account of his experiences with the Greenland ice sheet. Soon people began to understand that Agassiz's ideas were not so outrageous.

Agassiz immigrated to the United States in 1846, where he assumed a professorship at Harvard University and founded the university's Museum of Comparative Zoology. He spent the rest of his career at the museum, going on field trips throughout New England and New York, finding evidence that the ice ages had covered not only northern Europe but also the northern half of North America. At his death in 1873, the theory of ice ages was widely accepted worldwide, and a new generation of scholars was documenting evidence of glaciation, expanding the new field of glacial and ice age geology.

What Caused the Ice Age Cycles?

By the mid-twentieth century, studies of glacial deposits in Europe and North America revealed that there was no single ice age, but at least four or five ice ages on all the northern continents. In North America, they were known (from youngest to oldest) as Wisconsinan, Illinoian, Kansan, and Nebraskan. The names were derived from the southernmost state where the deposits of that event had been found. Europe had five events recorded by stacked glacial deposits: Würm, Riss, Mindel, Gunz, and Donau. For decades, disputes continued on how these events matched up until better dating methods showed that Riss and Würm were equivalent to parts of the Wisconsinan, Illinoian was equivalent to the Mindel, Kansan was the same as Gunz, and Nebraskan corresponded to the Donau. Even though the knowledge of the multiple episodes of glacials and interglacials was undergoing better documentation, there was still no clear evidence for what had caused glacial advances in the first place.

It is ironic that the solution to the problem had been worked on for decades but could not be tested with available geological evidence before 1975. Scottish physicist and mathematician James Croll was the first to propose a solution in 1864, while Agassiz's ideas were still new and controversial. Building on the work of French astronomer Urbain Leverrier, Croll argued that the amount of sunlight the earth receives (*insolation*) as controlled by the changes in the earth's orbit around the sun controlled the ice age cycles. The longest of these cycles was known as *eccentricity* (fig. 9.3A). The earth's orbit around the sun is not a perfect circle but an ellipse. Every 100,000–110,000 years, the shape of that ellipse changes from nearly circular to slightly oblong. When the orbit is circular and eccentricity is low, the earth passes closer to the sun in the winter part of the cycle, and there is more solar radiation on the poles, resulting in warming and melting of ice sheets. When the orbit is highly eccentric, the earth is much farther from the sun during the winter, and the ice accumulates.

Croll also calculated the existence of a second cycle, known as the *tilt*, or *obliquity* cycle (fig. 9.3A). Today, the earth's spin axis is tilted 23.5 degrees from the plane of its rotation around the sun, which creates seasons. When the North Pole tilts toward the sun, it is the Northern Hemisphere summer; when it tilts away from the sun, it is winter. This angle has not remained constant but fluctuates on a cycle with a duration of 41,000 years. In the past, the tilt is as steep as 24.5 degrees or as shallow as 21.5 degrees. When the tilt is steeper (24.5 degrees), the poles receive more solar radiation and the ice melts; when it is shallower, the poles receive less solar radiation and the ice accumulates.

The third cycle is *precession*, or "wobble," of the earth's axis. The spin axis of the earth wobbles like a spinning top that slowly loses momentum as it winds down (fig. 9.3A) so that the North Pole points to different stars over a 22,000-year cycle. Currently, the North Pole points to Polaris, the North Star, but halfway through that cycle (about 11,000 years ago), it pointed to Vega as the North Star. (Interestingly, this is not long before the Sumerians and

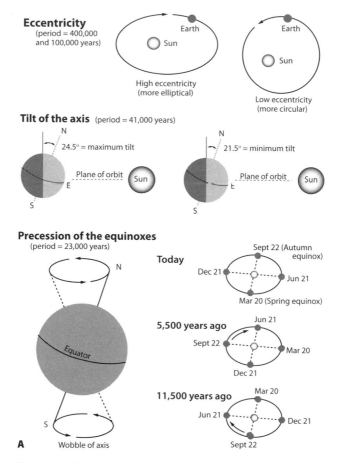

Fig. 9.3. A, Diagrams of the three main cycles of orbital variation. B (*opposite*), These cycles have three independent curves of insolation, which interfere and interact to form the complex jagged curve of observed climatic variation (*bottom*). (From Prothero and Dott 2009)

Babylonians invented astrology, so the skies and constellations no longer physically match what the zodiac claims—yet another reason why astrology is bunk.) Once again, times when the North Pole gets more sunlight in this cycle contribute to melting the ice caps.

These three cycles interact (fig. 9.3B) as three different wave functions, with different spacing of peaks and valleys. As sound waves can interfere to produce unexpected frequencies, so do these three cycles of orbital variations.

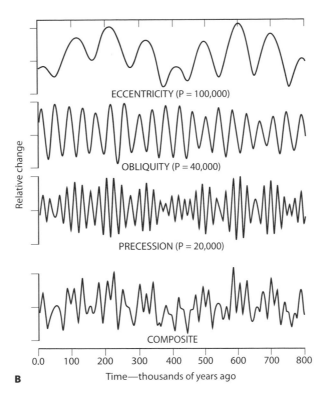

The composite curve (fig. 9.3B) thus has a complex, irregular jagged sawtooth shape.

When maximum solar insolation peaks, the three cycles coincide, enabling the prediction of peaks of melting and an interglacial interval. When the minimum solar radiation is predicted by all three cycles, there should be cooling and the peak of a glacial interval.

Croll's full theory was published in *Climate and Time* (1875), and he kept publishing his ideas until his death in 1890. The ability to date geologic events such as ice ages was still in its infancy in 1875; therefore, there was no way to determine whether his predictions were borne out by geologic evidence. The topic was considered an interesting but unproven speculation for decades until Milutin Milankovic, a Serbo-Croatian astronomer and mathematician, revived it. Beginning in 1907, he worked on dating geologic events, even when he was interrupted by three wars and even during his internment in an Austro-Hungarian camp during World War I. Milankovic updated and recalculated Croll's ideas, using recently improved data on the earth's orbit. He performed painstaking calculations by hand on paper (in an age long before computers made this job much easier). Milankovic realized that Croll was basically right, except that the crucial factor was not how much winter sun the poles received but how much summer sun they received. In 1930, he published *Mathematical Climatology and the Astronomical Theory of Climatic Changes*, which received more attention from the geological community than Croll's ideas. *Canon of Insolation and the Ice Age Problem* (1941) was still coming off the presses when the Nazis invaded Yugoslavia, and Milankovic had to flee as the printer's firm in Belgrade was destroyed. The final pages were reprinted elsewhere. After the war, he continued his work on this and other scientific problems (for example, he created the leap year rule for our modern calendar). He died in 1958 at age 79.

Solving the Mystery

Although tremendous progress had been made in ice age geology since Croll's time, there were still no reliable dates on the ice age cycles that could conclusively test the Croll-Milankovic hypothesis. An even bigger problem was that the record of ice ages on land was woefully incomplete. Only the largest ice advances had left permanent glacial deposits behind. Most of the smaller events left lesser deposits that had then been scoured away during the next major glacial advance.

The solution came from an unexpected source: the deep sea. Since the end of World War II, oceanographic institutions around the world had been sending out vessels to explore the unknown world that makes up 75 percent of the earth's surface. They collected long cores of deep-sea sediments from nearly every part of the ocean, and these cores turned out to have an almost continuous "rain" of sediments and planktonic microfossils that spanned millions of years with few gaps in the record. By the mid-1960s and 1970s, the DSDP (Deep Sea Drilling Project) had carefully collected and dated thou-

Fig. 9.4. The *Glomar Challenger*, the main research vessel of the Deep Sea Drilling Project, which drilled thousands of sediment cores all over the world's oceans and solved many geologic mysteries. (Photo courtesy NOAA)

sands of cores (fig. 9.4). Earth's climatic history was recorded in detail on the seafloor and could be studied on a millimeter-by-millimeter basis, allowing examination of events on nearly a yearly or decadal scale, even though these cycles occurred hundreds of thousands to millions of years ago.

In the early 1970s, the multimillion-dollar CLIMAP (CLImate MApping and Prediction) project, sponsored by the National Science Foundation, analyzed the cores in detail and could detect cycles of warming and cooling by the temperature tolerances of the planktonic microfossils found in the cores. They also developed a method of analyzing the isotopes of oxygen trapped in the shells of these plankton, which faithfully recorded the ancient seawater temperatures and ice volumes. By 1972, a detailed record documented at least 19 glacial-interglacial cycles in the past few million years, not just four or five that the terrestrial glacial deposits showed. The final stage was a mathematical method called *spectral analysis*, which broke down the complex composite climate curve into its three main component cycles. Sure enough, the analysis showed the 100,000-year eccentricity, the 41,000-year tilt, and the 22,000-year precession cycles predicted by Croll a century earlier. Orbital variations were indeed the "pacemaker" of the ice ages, forcing the planet to shift from glacial to interglacial modes. This landmark discovery was published in the famous 1976 pacemaker paper by Jim Hays, John Imbrie, and Nick Shackleton (1976). (Jim Hays was also on my dissertation committee and published several papers on radiolarians with me and his student Dave Lazarus.) Later retrospectives by science journalists regarded the pacemaker paper as one of the 10 most important scientific discoveries of the 1970s and, indeed, of the entire latter half of the twentieth century.

The study of Croll-Milankovic cycles has since blossomed, and these cycles have been found in deposits of many ages, so they are always an important part of earth's climate. Clearly, whenever the earth is covered in ice caps, it is highly sensitive to the amount of solar radiation that it receives. This raises a larger question: what causes the earth to have ice caps in the first place and to slip into an "icehouse" mode of climate? Only 100 million years ago, during the peak of the age of dinosaurs, the earth had a "greenhouse" climate, with no polar ice caps and huge oceans that drowned the continents (see Prothero 2009). The polar regions were covered in temperate vegetation and were inhabited by crocodiles, turtles, and dinosaurs that had to cope with six months of darkness (even though the temperature was balmy in winter). Why did the greenhouse of the dinosaurs vanish and our present-day icehouse replace it?

The story is complex and beyond the scope of this book, but the short answer is that by 33 million years ago, the polar ice caps were on Antarctica to stay and reached their present dimensions 15 million years ago. Many different ideas have been proposed to explain why this occurred, but the most widely accepted belief is that there has been a significant decrease in atmospheric carbon dioxide over the past 50 million years. Another possible contributor is the modern Circum-Antarctic Current (fig. 9.5), one of the largest currents in

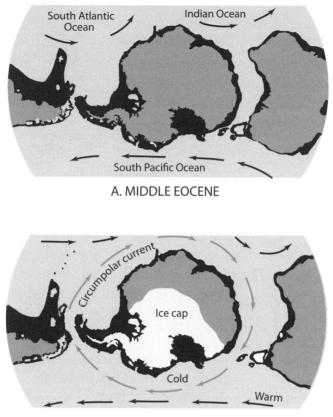

Fig. 9.5. The changes in the Circum-Antarctic current as Australia and South America separated from Antarctica. (From Prothero and Dott 2009; redrawn by Pat Linse)

the world ocean. It spirals around the Antarctic and stops warm water from reaching the region. This current acts like a "refrigerator door" locking in the cold air over the South Pole and allowing it to become glaciated. Since the mid-1970s, a number of scientists have argued that this current was first generated when Australia and South America pulled away from Antarctica 33 million years ago. This idea is still discussed and debated (see Prothero 2009, chap. 8), but there are strong coincidences between the timing of the opening of these oceanic gateways and the first appearance of this current as recorded in deep-sea cores.

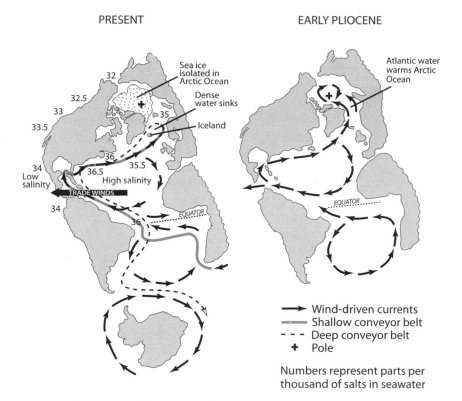

Fig. 9.6. Changing circulation as the closing of the Panama gap forces tropical waters from the Caribbean toward the Gulf Stream, rather than out to the Pacific. This warm current brought tropical moisture to the polar regions and allowed the Arctic ice cap to build up. (From Prothero and Dott, 2009; redrawn by Pat Linse)

If the Antarctic was glaciated 33 million years ago, when was the Arctic glaciated? Recent evidence shows that its present-day ice cap formed about 3 million to 4 million years ago. Since the 1970s, a number of scientists have pointed out that this growth of the Arctic ice cap coincides with the final closure of the Isthmus of Panama and the development of a land corridor along Central America. Before that time, warm ocean currents flowed through the Caribbean and out into the Pacific. Once they were blocked by Panama, they were forced to circulate northward to form the Gulf Stream, which brings warm moist waters from the tropics up into the North Atlantic (fig. 9.6). Today, this warm water from the Gulf Stream means that the British Isles have a mild and rainy climate, even though they are as far north as frozen Labrador. It also means that this tropical moisture is brought up to the polar regions, which were already cold. Once moisture is added to this freezing atmosphere, they produced snow and ice and generated the frozen wasteland of the Arctic, with its large floating ice sheet over the Arctic Ocean.

Ice Caps of a Supercontinent

Since the first good maps of the Atlantic Ocean were drawn in the 1500s, scholars such as Francis Bacon, Benjamin Franklin, and Alexander von Humboldt had noticed the match between the coastlines of western Africa and eastern South America and speculated that they might have once fit together. The idea of drifting continents remained dormant until the early twentieth century when a brilliant young German meteorologist named Alfred Wegener proposed that the earth's continental landmasses had once been in a supercontinent that he called *Pangea*, or "all earth," and had since drifted apart. In his 1915 book *Die Entstehung der Kontinente und Ozeane* (*The Origin of Continents and Oceans*), Wegener amassed nearly all the evidence that was available at the time, from the fit of western Africa against the South American coast, to the similarity of fossils and deposits between Africa and South America (and sometimes other southern continents, including India, Australia, and Antarctica) about 250–300 million years ago, to the striking way in which the ancient basement rocks of South America looked as though they once joined those of West Africa and had since ripped apart by the opening of the Atlantic.

This evidence of continental movement should have been convincing to geologists back then, but they rejected, ignored, or ridiculed Wegener and his ideas, primarily because Wegener was a meteorologist with limited geologic training; the geological community regarded him as an "amateur" or "outsider." They did not trust his data, given his limited background in rocks and fossils. The bigger difficulty remained the mechanism for how continents might drift around the globe. Why was there no evidence of huge masses of crumpled-up land on the front edges of continents as they plowed across the oceans? These questions would not be answered until revolutions in marine geology and paleomagnetism came along in the 1950s to show that Wegener's critics were wrong about many of their assumptions about the earth's crust. It turned out that oceanic crust is not thick nor long-lasting, but is easily subducted and thus does not usually end up like a crumpled carpet on the leading edge of a moving continent. Unfortunately, Wegener died in 1931, frozen to death during one of his expeditions to the Greenland ice sheet. He never lived to see his vindication 30 years later.

One of Wegener's powerful lines of evidence, however, came from his background in climates and meteorology. When he looked at maps of climatically sensitive rock types (fig. 9.7), he found that tropical coal swamps, subtropical deserts, and other evidence of climate about 250–300 million years ago made no sense when plotted on a modern map of the world, but fit together nicely at the expected latitudes when plotted on his map of Pangea. The most striking of all these climatically sensitive deposits was the evidence of huge glaciers that had once covered the South Pole among the various continents that made up the southern half of Pangea, known as Gondwanaland. Thick glacial deposits match up perfectly between South America and southern Africa, and some are found on the rest of the southern continents, such as India. If you put the continents in their Gondwana configuration, the ice sheet is a single simple polar mass. However, if you try to plot it on a modern map of the world, it would require a southern ice sheet that crosses the equator to India and covers some but not all of the South Atlantic and Southern Ocean but not the southern Pacific Ocean. The final clinching evidence comes from the deep, straight parallel scratches, or striations, in the bedrock formed as the rocks embedded in the glacial ice mass were rasped across the bedrock like a rake on the sand (fig. 9.8). In many cases, those scratches start in western Africa and can be matched up in a straight line with comparable scratches in Brazil. Either the glaciers plunged into the ocean and crossed the Atlantic in a straight line before jumping back out again to scratch Brazilian bedrock (clearly an absurdity), or there was no Atlantic Ocean when the glaciers scraped the ground from West Africa to Brazil.

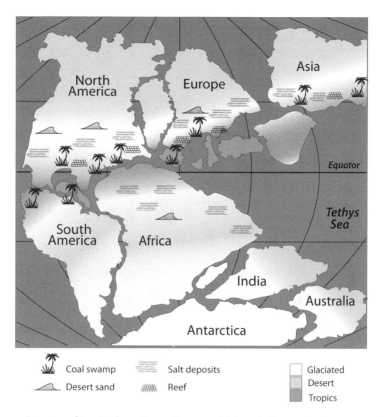

Fig. 9.7. Permian climatic deposits on Pangea, which only line up in the context of the Pangea continental configuration but make no sense in the framework of modern continental positions. (From Prothero and Dott 2009; redrawn by Pat Linse)

Putting this all together, we can map a huge glacial ice sheet centered largely over the ancient South Pole (then located in southern Africa), with portions covering southern South America and parts of India, Madagascar, Australia, and even Antarctica (fig. 9.8A). This continental glacier system apparently waxed and waned every 100,000 years according to the Croll-Milankovic cycles, which were just as effective 300 million years ago as they have been over the past 2 million years. The fluctuation of ice volume meant that during interglacials the seas rose dramatically as the ice melted and flooded the ocean basins and continents, while during glacials, the ice locked up a lot of water and the sea levels dropped. The discovery of these 100,000-year glacial cycles

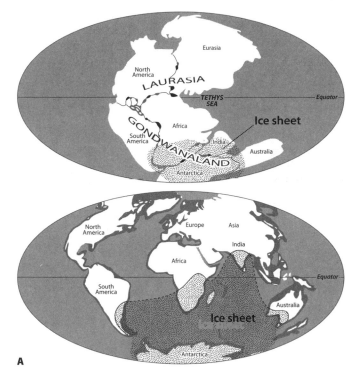

Fig. 9.8. A, Distribution of Permian glacial deposits in a Pangea configuration of continents (*top map*), compared to the nonsensical distribution in the modern arrangement of continents (*bottom map*). B, Glacial scratches from the Permian of South Africa, just below the Dwyka tillite, a thick glacial deposit. (From Prothero and Dott 2009; a redrawn by Pat Linse)

on Gondwanaland led to the solution of another important mystery: the cyclicity of nearshore deposits across the world in the Carboniferous, known as the "cyclothems." These often showed 20–30 fluctuations in sea level, and during their cycles, they built huge swampy nearshore deposits that entombed most of the coal deposits known around the world.

Ironically, geologists now believe that those same coal deposits are at least partially responsible for the transformation from a greenhouse world between 550 million and 300 million years ago, and the icehouse world of the Carboniferous and Permian, 300–250 million years ago. When these gigantic swamps developed, they absorbed huge amounts of carbon dioxide in their decaying plant material, which were then entombed in the earth's crust as coal. This in turn reduced global carbon dioxide levels in the atmosphere and produced an icehouse world. As a final twist, burning these same coals releases carbon dioxide into the atmosphere, which is a main contributing factor to global warming. What the earth took out of the atmosphere and locked up in the crust for more than 250 million years is being put back into the atmosphere over less than two centuries.

Snowball Earth

In the early 1960s, geologist Brian Harland at Cambridge University was working on an interesting problem. He had been studying glaciers and glacial deposits nearly all his life and discovered evidence of glacial deposits nearly all over the world dated to about 600 million years ago, before the "Cambrian explosion" that led to the diversification of multicellular animals in the world's oceans. In 1964, he published a paper arguing that, in addition to the global ice ages in the past 2 million years, and the Carboniferous-Permian ice age 300–250 million years ago, there was a third global ice age, around 700–600 million years ago. Dating these deposits was pretty imprecise in the early 1960s. There was still no way to determine how all the ancient continents fit together or at what latitude some of these glacial deposits had formed, but they had occurred nearly every place on earth that had deposits of this age. However, it was clear that at least some of these glacial sediments had been formed in sea-level glaciers near the equator because they were underlain and overlain by limestones, which today are restricted to shallow marine environments of tropical and subtropical latitudes (plate 14). The implications are staggering. Today, the only tropical glaciers occur in high mountains, like the Andes or Kilimanjaro. If there were glaciers in the tropics *at sea level* 600 million years ago, then the planet would have had major ice sheets, even on the equator, which would imply that the earth was nearly frozen over.

At the same time that Harland was working on the problem geologically, geophysicist Mikhail Budyko at the Leningrad Geological Observatory was working on simple climatic models that simulated what it would take for the earth to go from an ice-free state to a frozen ice ball. It turned out that it took very little difference in solar energy for the planet to make that transition, because the reflectivity of ice (its albedo) would bounce back more and more sunlight as the ice sheet grew in a feedback loop to cool the planet even further, creating more ice. Eventually, it would reach a frozen state. Budyko had shown that it was easy to freeze the planet into a snowball, but once it was frozen over, it seemed impossible to warm it up again. Since Budyko's experiments, it has been shown that the sun was dimmer and produced 6 percent less energy 600 million years ago than it does today, making a frozen snowball planet more likely. Mechanisms such as the weathering of rocks would soak up the carbon dioxide in the atmosphere and start the planet toward glaciation.

Neither scientist nor their ideas had an effect for a while, because there was not enough data to determine at what latitude the glacial deposits had formed and whether glacial deposits were all precisely the same age. Then, in 1992, my friend Joe Kirschvink at Caltech proposed a radical idea. Joe is a genius whose skills extend across paleomagnetism, geobiology, geochemistry, and paleoclimates; he builds his own lab equipment and writes his own computer code. Joe has more original (and sometimes outrageous and provocative) ideas than any scientist I know. He used the newly obtained paleomagnetic data analyzed in his lab to decipher the true positions of the continents 700 million years ago and to show that indeed many of the glacial sequences were tropical or subtropical (plate 14). His paleomagnetic studies also provided a much finer age control, so he could demonstrate that glaciers were indeed the same age worldwide. Finally, his paper concluded with a solution to Budyko's dilemma. If the planet were covered by ice as a "snowball earth," eventually the carbon dioxide and other greenhouse gases released by volcanoes under and poking through the ice would build up enough to warm the planet and quickly thaw the snowball. Joe also found other supportive evidence from banded iron deposits of the same age, which were consistent with an oxygen-poor ocean trapping dissolved iron under the ice caps.

Joe's original 1992 paper was published without fanfare, probably because it was only two pages long and was buried in an expensive 1,400-page volume about early life. It went unnoticed for a while. Joe had put his findings out there as an interesting idea worth exploring, and then he moved on to the many other intriguing topics he was constantly working on. Eventually, the idea of a snowball earth came to the attention of Harvard geologists Paul Hoffman, Daniel Schrag, and their colleagues, who realized that it explained many of the peculiar features of ancient glacial deposits they found in northern Canada and in Namibia. In the late 1990s, Hoffman and Schrag gave many talks and published many high-profile papers on the subject, and soon it was one of the hottest topics in geology.

Hoffman and Schrag and others have pointed to the peculiar feature of limestones (now turned into magnesium-rich carbonate rocks called dolostones) that lie above the glacial deposits. These "cap carbonates" have many interesting geologic structures and odd chemical properties that suggest they were crystallized directly out of seawater without the usual algae and animal shells helping them along. Geochemists have modeled the conditions that would produce these weird rocks, and most of the models involve releasing a huge amount of trapped carbon from the ocean basins, possibly as methane surrounded by ice, once the volcanic greenhouse gases break the icy snowball grip on the planet.

Naturally, the idea of an earth frozen over for millions of years, then switching to a runaway greenhouse, has been controversial among geologists. Although there are many arguments about how strongly particular outcrops support the snowball model, there are some that are beyond dispute. The best studied is the Elatina deposits in southeastern Australia, where there is a clear sequence of limestone–glacial deposit–limestone, and the strata have excellent dating and undisputed paleomagnetic evidence that they were once right on the equator.

There are some geologists who prefer a less extreme version, the "slushball earth," with much of the earth nearly frozen but some areas still above freezing. There are always disputes about individual data points in paleomagnetism and geochemistry and whether rocks identified as glacial meet the strictest criteria. Nevertheless, most geologists familiar with the problem agree that the planet came close to freezing over about 700–600 million years ago. Whatever the details, there was a significant extinction among the marine algae, known as *acritarchs*, that had lived a long time before the snowball. Once the snowball episode was over, we see the first evidence of multicellular soft-bodied animals in the Ediacaran fauna, the first megascopic evidence of life in the entire geologic record.

Finally, there is also good evidence of yet another snowball episode more than 2 billion years ago. Although the exposures are less widespread and studied, nonetheless there are outstanding glacial deposits (like the Gowganda sequence of Ontario) associated with limestones and sea level, and whose paleomagnetic data clearly place them in equatorial latitudes.

There's an interesting coda to this issue. In the 1990s, the Martian meteorite found in the Allan Hills of Antarctica was all the rage as numerous scientists argued about whether it showed evidence of Martian microorganisms. Although this controversy has not been resolved, it points to an interesting paradox: Mars is now a frozen snowball planet with its water locked up in ice, but it might have had liquid water and life long ago. If the earth had frozen over completely 700–600 million years ago and never recovered, we would have looked much like Mars—and life on this planet would be single celled and not much more sophisticated than the possible fossils in the Martian meteorite. It's an interesting thought experiment. What if the snowball earth had not melted? Would multicellular life have developed, or would we look like the frozen Mars with only traces of simple life from its warmer past?

Thus, we now know of at least four major "icehouse" glacial periods in earth history: 2 billion years ago, 700–600 million years ago, 300–250 million years ago, and the last 33 million years. Some people have argued that these "icehouse" events are part of a cycle, but with only four of them (not that evenly spaced out), it is hard to make that case (fig. 10.3). We will consider the possibility of icehouse-greenhouse cycles in the next chapter.

Past and Future Glaciations?

Picture any spot in northern North America 20,000 years ago, during the peak of the last glaciation. If you stood in Boston, New York City, or Chicago, there would have been a mile of ice above you, comparable to the thickness of the Antarctic ice cap. The edge of the glaciers extended all the way to the midcontinent (fig. 9.9A). The areas to the south of the glaciers (the central and southern Plains, southeastern United States) would have been tundra or steppe vegetation or the northern conifer forests like those in Canada, not the dry grasslands or pine forests now found in the Great Plains or Deep South (fig. 9.9B). Only Florida would have looked similar because it retained its subtropical climate during both glacial and interglacial intervals. Farther west, the dry deserts of Nevada and western Utah were wet and full of lakes that filled every basin between the ranges. Some huge lakes included Glacial Lake Missoula mentioned in chapter 5 or the gigantic Lake Bonneville, whose dried remnants are the "Great" Salt Lake and the Bonneville Salt Flats. Both lakes drained catastrophically to the Pacific Ocean through the Columbia and Snake rivers on more than one occasion. The dry chaparral of Los Angeles was cooler and moister, with numerous pine trees, rather than the semidesert scrub that now covers the region.

The same would have been true in Eurasia. All of Scandinavia, Great Britain, most of Europe north of the Alps, and most of Siberia were under a thick sheet of ice. The high mountain ranges of the world—the Himalayas, Andes,

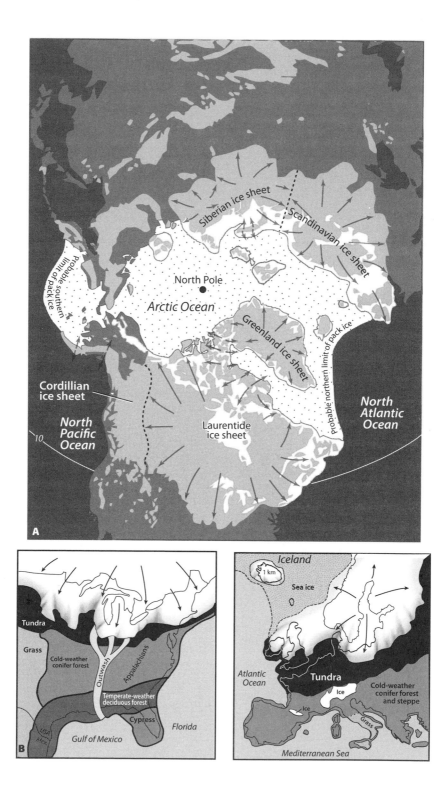

Siberian ice sheet

Scandinavian ice sheet

probable southern limit of pack ice

North Pole

Arctic Ocean

Greenland ice sheet

Probable northern limit of pack ice

Cordillian ice sheet

North Pacific Ocean

10

Laurentide ice sheet

North Atlantic Ocean

A

Tundra

Grass

Cold-weather conifer forest

Outwash

Appalachians

Temperate-weather deciduous forest

Cypress

Florida

USA
Mex.

Gulf of Mexico

B

Iceland

1 km

Sea ice

Atlantic Ocean

Tundra

Ice

Cold-weather conifer forest and steppe

Grass

Ice

Mediterranean Sea

Alps, Rocky Mountains, and the rest—would have had much larger valley glaciers than they do now, and their glaciers would have spread for miles around them. The Arctic Ocean expanded into the Bering Sea and much of the North Atlantic south of Iceland. The Antarctic ice caps also expanded until nearly all the Southern Ocean was frozen over. There would have been solid ice linking the Antarctic Peninsula to the tip of South America, and the floating ice shelf would have reached almost to Australia and South Africa as well.

All that glacial ice pulled an enormous amount of water from the ocean basin. Consequently, sea level dropped more than 120 m (390 feet) below its present level, so the entire continental shelf regions of most of the world's ocean would have been exposed land, with low, flat floodplains and broad river valleys. We know this not only from evidence of sediments on the continental shelf well below sea level but also from mammoth teeth dredged up from water more than 100 m deep near the edge of the continental shelf. Broad land bridges allowed animals to walk between Siberia and Alaska. Former islands, such as Great Britain, the Balearic Islands, including Majorca, Corsica, Sicily, Sardinia, Malta, Madagascar, and the Indonesian Archipelago, were connected to the mainland and part of their adjacent continents so that animals walked freely back and forth.

This was the world of the ice ages, which lasted for more than 2 million years and last peaked 20,000 years ago. Our present-day interglacial "warm" episode, known as the Holocene, has been under way for only the past 10,000 years. It is merely another of many interglacial episodes between glacial cycles. Technically, we are still in the "ice ages," but we're getting an interglacial reprieve that happens to span all of human history and a big chunk of our prehistory.

If the Croll-Milankovic cycles are still affecting the planet (and there's no doubt that they are), what can we predict for the future of our planet? How long do we have before the next ice advance and a new ice age begins? We can

Fig. 9.9. (*opposite*) A, North polar view of the ice cover at the peak glaciation 20,000 years ago. B, Map of the peak glaciation and the distribution of vegetational belts 20,000 years ago. (After Prothero and Dott 2009)

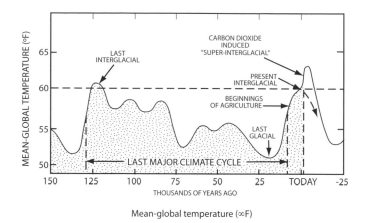

Fig. 9.10. The last 150,000 years of glacial-interglacial cycles, showing the typical 10,000-year duration of the previous interglacial 125,000 years ago and the predicted end of our current interglacial after 10,000 years. (After Prothero and Dott 2009; redrawn by Pat Linse)

look back 125,000 years ago at the previous interglacial episode. That penultimate interglacial, and all the ones before it, lasted approximately 10,000 years before glaciation returned (fig. 9.10). We have just gone through 10,000 years of the current Holocene interglacial, so if the normal orbital variation control of ice ages were the only important factor, we should be headed into the next ice age any time now. When that happens, most of our northern continents will again be covered in ice (plate 14B).

This assumes that nature will take its normal course. If human-induced global warming overwhelms the effects of global solar radiation, then instead of gradually sliding into the next ice age, we will go to the other extreme: a greenhouse world that would be "super-interglacial" (fig. 9.10). Some people might prefer this climate to having a mile of ice above them, but as we shall see in the next chapter, a greenhouse world has severe drawbacks.

FOR FURTHER READING

Agassiz, L. 1840. *Etudes sur les glaciers*. Agassiz, Jent & Glassman, Neuchatel.

Bolles, E. B. 1999. *The Ice Finders: How a Poet, A Professor, and A Politician Discovered the Ice Age*. Counterpoint, New York.

Budyko, M. I. 1969. The effect of solar radiation variations on the climate of the earth. *Tellus* 21:611–19.

Fagan, B. M. 2009. *The Complete Ice Age: How Climate Change Shaped the World.* Thames & Hudson, London.

Flint, R. F. 1971. *Glacial and Pleistocene Geology.* Wiley, New York.

Gribbin, J. 2003. *Ice Age: How a Change of Climate Made Us Human.* Penguin, London.

Harland, W. B., and M. J. S. Rudwick. 1964. The great Infra-Cambrian glaciation. *Scientific American* 211 (2): 28–36.

Hays, J. D., J. Imbrie, and N. J. Shackleton. 1976. Variations in the earth's orbit: pacemaker of the ice ages. *Science* 194 (4270): 1121–32.

Hoffman, P. F., A. J. Kaufman, G. P. Halverson, and D. P. Schrag. 1998. A Neoproterozoic snowball earth. *Science* 281:1342–46.

Imbrie, J., and K. P. Imbrie. 1979. *Ice Ages: Solving the Mystery.* Enslow, Short Hills, NJ.

Kirschvink, J. L. 1992. Late Proterozoic low-latitude global glaciation: the Snowball Earth. *In* J. W. Schopf and C. Klein, eds., *The Proterozoic Biosphere*, 51–52. Cambridge University Press, Cambridge.

Levenson, T. 1989. *Ice Time: Climate Science and Life on Earth.* Harper & Row, New York.

Matsch, C. L. 1976. *North America and the Last Great Ice Age.* McGraw-Hill, New York.

McDougall, D. 2006. *Frozen Earth: The Once and Future Story of Ice Ages.* University of California Press, Berkeley.

Muller, R. A., and G. J. MacDonald. 2002. *Ice Ages and Astronomical Causes.* Springer, Berlin.

Pielou, E. C. 1991. *After the Ice Age: The Return of Life to Glaciated North America.* University of Chicago Press, Chicago.

Prothero, D. R. 2006. *After the Dinosaurs: The Age of Mammals.* Indiana University Press, Bloomington.

Prothero, D. R. 2009. *Greenhouse of the Dinosaurs.* Columbia University Press, New York.

Prothero, D. R., and R. H. Dott, Jr. 2009. *Evolution of the Earth.* 8th ed. WCB/McGraw-Hill, New York.

Walker, G. 2003. *Snowball Earth.* Bloomsbury, Philadelphia.

10

Greenhouse Planet

Too Hot to Handle?

[Carl] Sagan called [the earth] a pale blue dot and noted that everything
that has ever happened in all of human history has happened on that tiny pixel.
All the triumphs and tragedies. All the wars. All the famines. It is our only home.
And that is what is at stake—to have a future as a civilization. I believe
this is a moral issue. It is our time to rise again to secure our future.
—Al Gore, *An Inconvenient Truth*, 2006

Cretaceous Park

Imagine taking a time machine to the final half of the age of dinosaurs. Known
as the Cretaceous Period, it spanned the interval of time from 145 million
years ago until 65 million years ago, a total of 80 million years. This is 15 million years longer than the entire duration of the "age of mammals," or Cenozoic era, which has lasted for the past 65 million years. If you stepped out of
the time machine, you would not recognize much of the landscape and not
just because dinosaurs ruled the world. The Rocky Mountains did not exist;
most of California, Oregon, and Washington did not exist; and the Appalachians were not nearly so deeply eroded as they are today. The most notice-

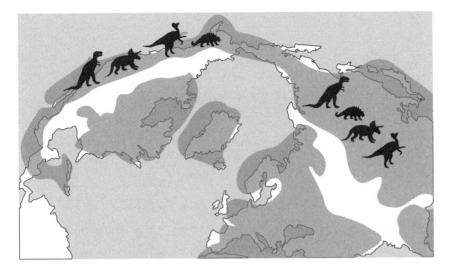

Fig. 10.1. Map showing the high sea levels of the Cretaceous that drowned the continents in many parts of the world and made it easier for dinosaurs from Montana to walk to Mongolia than to New Jersey. (Original drawing by Pat Linse)

able difference would have been a huge shallow seaway filled with predatory marine reptiles, such as mosasaurs, plesiosaurs, ichthyosaurs, and sea turtles, as well as gigantic fish, with huge flying reptiles soaring overhead. Today, if you go to western Kansas or South Dakota, you find the ancient marine beds filled with these marine reptile fossils. The Western Interior Cretaceous Seaway (fig. 10.1) once extended all the way from the Arctic Ocean to the Gulf of Mexico, flooding the entire Great Plains and much of the future Rocky Mountain region, cutting North America in half. Although it was only a few hundred meters deep at its deepest, it was an impassable barrier for land animals. The dinosaurs of Montana, Alberta, or New Mexico, which lived on the western shore of this seaway, had more in common with the dinosaurs of Mongolia than with the dinosaurs of New Jersey because there was a temperate-forest land bridge over the Bering Strait to Asia.

That would be the second striking fact about the Cretaceous world: instead of polar ice caps, the Arctic and Antarctic were temperate or even subtropical, with abundant trees, and a large fauna of dinosaurs, turtles, crocodiles, frogs, salamanders, and other animals living there. Many animals, such as crocodiles, cannot tolerate even the slightest freezing temperature, so they must have lived in relatively mild conditions even though above the Arctic and Antarctic circles the winters mean six months of darkness. For example, Cretaceous rocks on the North Slope of Alaska (below freezing most of the year today) yield abundant fossils of bald cypress, cycads, ginkgoes, ferns, and many other temperate and subtropical plants typically found in the swamps of Georgia or Florida today. Instead of the subfreezing annual temperatures, the summer temperatures in the region were about 10°C (50°F), or cool temperate.

This is also true of the southern polar regions. Decades of work in southern Australia and even in Antarctica show the region inside the Antarctic Circle supported a varied fauna of dinosaurs, fish, turtles, flying pterosaurs, birds, amphibians, and even small mammals. The temperatures were similar to Alaska at the time, with lush, green vegetation. It included ferns, ginkgoes, cycads, podocarps ("yew pines"), numerous flowering plants, and *Araucaria* trees (known today as the Norfolk Island pine, or the monkey-puzzle tree). As in the Arctic, these plants could tolerate six months of darkness but not freezing, so they would have been dormant about half the year. Dinosaurs show an interesting mixture of ecologies. Most dinosaurs from Alaska are the same as the duckbill dinosaurs in southern Alberta, so it is reasonable to assume that they migrated from the region in herds during the winter darkness and returned in the summer (as most Arctic animals do today). Many dinosaurs from southern Australia, however, are small bodied and could not migrate. Their bones have clear growth lines, suggesting dormancy or hibernation half the year. Other Australian Cretaceous dinosaurs show no such growth lines and have large eye sockets, which support the notion that they were active year round and used their large eyes to see in the dim world of polar winter.

How could these animals and plants survive six months of darkness each year? Temperatures had to be mild, because so many of these organisms are intolerant of freezing conditions. Most paleoceanographers agree that polar regions must have been bathed in oceanic currents coming up from the tropics, which kept the climate mild and minimized the temperature difference between the poles and the equator. In addition to the absence of ice anywhere on the planet, melting polar ice caps would help explain why sea levels rose and drowned the continents, forming the Western Interior Seaway (fig. 10.1). Finally, paleoclimatologists have found numerous lines of evidence indicating atmospheric carbon dioxide must have been about 2,000 ppm (parts per million), almost 10 times the pre–Industrial Revolution values of 270 ppm. This was a true "greenhouse planet," with no ice or polar ice caps, high sea levels, and very high levels of greenhouse gases like carbon dioxide in the atmosphere.

Where did all this carbon dioxide come from? A number of potential sources have been discovered by geologists over the past decades of research. One factor was that the supercontinent of Pangea (fig. 9.7) was breaking up rapidly during the Cretaceous. The ocean floor was spreading apart at rates never seen again in earth history as the continents raced away from one another. Seafloor spreading brings up huge volumes of volcanic gases from the earth's mantle, among which are abundant greenhouse gases such as carbon dioxide. Another contributing factor are huge lava eruptions straight from "hot spots" or plumes of molten material in the mantle. Most of these eruptions occurred beneath the ocean, piling up huge volumes of basalts in submarine plateaus such as the Ontong Java Plateau in the western Pacific. This plateau (one of many now known) consists of 1.5 million cubic kilometers (360,000 cubic miles) of lava erupted in less than a 1-million-year interval about 122 million years ago. Other similar plateaus include the Hess Plateau, the Kerguelen Plateau, and the Shatsky Rise, all of which would have poured immense amounts of carbon dioxide into the Cretaceous atmosphere. There is no problem with producing the Cretaceous greenhouse atmosphere given the ongoing geologic events. The problems remain in the details of when the carbon dioxide erupted, and how these events match the changes in sea level that occurred throughout the Cretaceous.

This "greenhouse of the dinosaurs" reached an even hotter level in the early Eocene, about 55 million years ago. At that time, a huge "burp" of methane was released from ice crystals embedded in oceanic sediment, and suddenly global temperatures rose higher than they had been for 300 million years. For example, in the Canadian Arctic are plant fossils that suggest even warmer conditions than in the Cretaceous. These plants include sequoias, dawn redwoods, elms, oaks, *Liquidambar*, ginkgoes, *Viburnum*, and bald cypresses. On the basis of modern forest analogues, the mean annual temperature in the Canadian Arctic was as high as 25°C (77°F), or temperate to subtropical. Living in those forests were abundant crocodilians, pond turtles, amphibians, monitor lizards, garfish, as well as primitive horses, tapirs, rhinoceroses, lemur-like primates, rodents, and many other creatures that required a dense forest with ponds and subtropical conditions to survive.

Greenhouse-Icehouse Cycles?

This greenhouse world came to a slow end throughout the 21 million years of the Eocene (55–34 million years ago). It climaxed with the appearance of glacial ice caps on Antarctica 33 million years ago, which marked the beginning of the modern icehouse world. As discussed in chapter 9, the end of the "greenhouse of the dinosaurs" is a complex story, which certainly involves some mechanism of lowering carbon dioxide in the atmosphere and may be a function of changes in oceanic circulation and the development of the Circum-Antarctic Current (fig. 9.5). The full details of this story are given at length in my book, *Greenhouse of the Dinosaurs* (Prothero 2009, chap. 8).

Was the "greenhouse of the dinosaurs" a unique event in geologic history? No! If you travel across the Midwest and many parts of the Rockies and the Colorado Plateau, almost every Paleozoic outcrop and road cut consist of fossiliferous marine rocks produced in shallow tropical seas that covered the planet from 600 million to 320 million years ago (fig. 10.2). These include the fossiliferous marine shales and limestones around Cincinnati, Ohio; the incredibly fossiliferous limestone quarries of Indiana, Illinois, and Iowa; the great limestones of Kentucky from which Mammoth Cave was etched—or similar limestones like the Redwall limestone in the Grand Canyon, the Madison limestone in the northern Rockies, or the Pahasapa limestone in the Black Hills. All are products of another huge tropical seaway that drowned the entire

Fig. 10.2. Typical example of fossiliferous limestones from the Midwest, showing the continent was drowned during most of the Paleozoic by enormous shallow tropical seas, something like the Bahamas but on a continental scale. A, Early Carboniferous limestone from Illinois, made entirely of the pieces of crinoids, or "sea lilies." It is estimated that several trillion of these animals made up units such as the Burlington Limestone. B, An Ordovician outcrop from near Cincinnati, with two large coral heads growing one on top of the other. (Photos by the author)

continent except for small areas (like the Appalachian Mountains) that emerged from the waves. During this time span, there were few or no glacial ice caps on the polar continents, and carbon dioxide must have been at levels comparable to the 2,000 ppm of the Cretaceous.

Thus, we have greenhouse worlds between the Cambrian and early Carboniferous (600–320 million years ago) and in the later Mesozoic to Eocene (200–33 million years ago), and icehouse worlds of the "snowball earth" (700–600 million years ago), the great Permian Gondwana ice sheet (fig. 9.7) of 320–200 million years ago, and the current icehouse world from 33 million years ago until now (fig. 10.3). Three icehouses separated by two greenhouses over roughly 500–600 million years. Some geologists calculate a roughly 250–million-year-long greenhouse-icehouse cycle, and many suggestions have been proposed to explain this apparent cyclicity. However, it is difficult to make the case that there is a true alternation of two stable states here, because there are only 2.5 "cycles" in all, not enough events to make the case statistically. In addition, it does not appear that the causes of each transition are comparable or even similar. For example, the greenhouse-icehouse transition that occurred 320 million years ago was probably triggered by carbon dioxide trapped in huge coal deposits (see chapter 9), but no such similar event took place 33 million years ago during the last greenhouse-icehouse transition. Arguments about the possible causes of these events continue, but whatever direction the discussion takes, it is clear that the earth has had long periods of greenhouse climates with no polar ice caps, high carbon dioxide levels in the atmosphere, and sea levels that drowned most of the lower elevations of the continents.

The "Super-Interglacial" Greenhouse

Looking back at past greenhouse worlds gives us a different perspective on the global climate change our planet is experiencing. As we saw in chapter 9 (fig. 9.10), if the normal glacial-interglacial cycles of our current icehouse world were allowed to operate naturally, we should have begun heading into the next glacial cycle any time now. As we shall see later, the opposite is the case. The planet is already measurably warmer than it has been in any interglacial

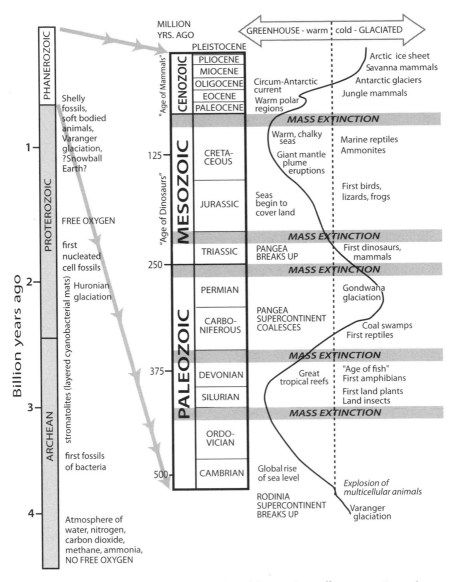

Fig. 10.3. The "greenhouse-icehouse" cycles of the past 600 million years. (Image by author; redrawn by Pat Linse)

in the past 2 million years, with carbon dioxide levels at 400 ppm and headed up past 600 ppm, higher than at any time in the past 5 million years. In the discussion that follows, we shall see some of the many effects of this "super-interglacial" greenhouse world we are inflicting on the planet.

How did scientists discover that the planet was undergoing global warming? As early as 1896, the Nobel Prize–winning Swedish chemist Svante Arrhenius showed that carbon dioxide was a powerful greenhouse gas. He realized that when the earth receives energy through solar radiation it comes in short wavelengths (mostly visible light and ultraviolet) that penetrate any atmospheric gases. Once the solar radiation is absorbed by the earth, the energy radiates back out as heat (infrared radiation), a long-wavelength form of energy blocked by greenhouse gases. Unable to radiate back into space, heat accumulates in the earth's atmosphere and contributes to global warming. Thus, the greenhouse gases act like the glass ceiling of a gardener's greenhouse, locking in heat but letting the light through. In 1896, Arrhenius calculated that doubling the level of atmospheric carbon dioxide would cause global temperatures to rise by 5°–6°C. This is remarkably close to the current estimates of scientists in the Intergovernmental Panel on Climate Change (IPCC) report in 2007, published more than a century later with more data and better instrumentation than was available to Arrhenius.

Atmospheric scientists occasionally discussed global warming and the greenhouse effect over the next 60 years, but no real systematic experiments to measure its effects were undertaken until 1958. Then Charles D. Keeling began his first measurements of atmospheric carbon dioxide that continue today, a half-century later. Keeling was trained as a geochemist at the University of Illinois and Northwestern University and then as a postdoctoral student at Caltech, where he invented one of the first devices for measuring carbon dioxide in the atmosphere. In 1956, he was hired at the Scripps Institute of Oceanography near San Diego to begin work on measuring changes in the earth's atmospheric carbon dioxide. Spurred by the legendary geologist Roger Revelle, Keeling used funding from the famous International Geophysical Year program to begin his research. In 1958, he began measuring both in Ant-

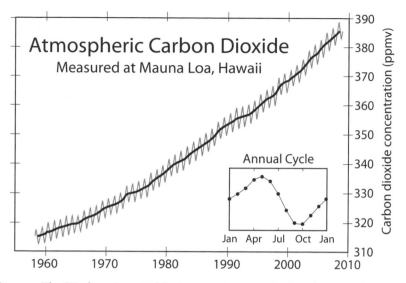

Fig. 10.4. The "Keeling Curve" of the increase in atmospheric carbon dioxide since 1958. (Modified from image at Globalwarmingart.com; redrawn by Pat Linse)

arctica and atop Mauna Loa, an extinct volcano in Hawaii, as well as a few other places. Most of these places were chosen because they were far from the effects of local cities or large forests and continents, so they gave the best possible averages for the global atmosphere.

By 1960, the second year of his experiment, he could already see the mean values climbing dramatically (fig. 10.4). After a few years, his funding from the National Science Foundation ended because his results were considered "routine." They were uninterested in supporting any long-term research that might take more than three to five years. This is typical of the vagaries of science funding in the United States, where safe short-term projects get all the regular funding, but long-term projects, or more daring projects that don't guarantee success have no chance at money. Ironically, after the National Science Foundation cut his funding, they still used his results to warn the government about the dangers of global warming.

Keeling had to abandon his expensive trips to Antarctica, and he and Revelle scraped up enough money each year to keep the Mauna Loa Observatory going—and it is still collecting data today, more than 52 years later, making it the longest continuously collected data set on global climate anywhere in the world. Today, the Mauna Loa Observatory is a large facility, with numerous buildings and different experiments, well funded by the National Oceanic and Atmospheric Administration. It is an important landmark in the history of science, just as Mount Wilson Observatory in the mountains above Pasadena is the birthplace of the discovery of the expanding universe and the "Big Bang" theory.

Keeling's experiments generated the famous "Keeling Curve" showing the increase in carbon dioxide since 1958 (fig. 10.4). As already mentioned, it shows a dramatically increasing trend, year by year. There are also seasonal up-and-down cycles of carbon dioxide caused when Northern Hemisphere plants absorb the gas in the spring and then release it when they die back in the fall and winter. Such a monotonic change should have already been evident, but it always helps to look over longer time frames to see whether the increase is part of a "natural variation." This has been done on many different timescales. For example, in the famous temperature plot (fig. 10.5) of the past 1,000 years of temperature data (Mann et al. 1999), the past 900-plus years are reconstructed from tree rings, ice cores, corals, and other climatic indicators. The final century of data points comes from direct measurements. No matter which data sets are used the results are similar. Climate was stable within a narrow band through the past 1,000, 2,000, or even 10,000 years of the Holocene, with just slight warming events during the Climatic Optimum about 7,000 years ago, and the slight cooling of the Little Ice Age from the 1700s and 1800s. But the magnitude and rapidity of the warming represented by the past 200 years is simply unprecedented in all of human history—and it exactly coincides with the Industrial Revolution, when humans first began massive deforestation and burning coal, gas, and oil in large quantities, which released huge amounts of carbon dioxide. This is the clearest possible "smoking gun" for human culpability for global warming because no other explanation comes close to making sense of it. The curve is now legendary and known by the nickname "the hockey stick curve" because of its long, straight "shaft" of no

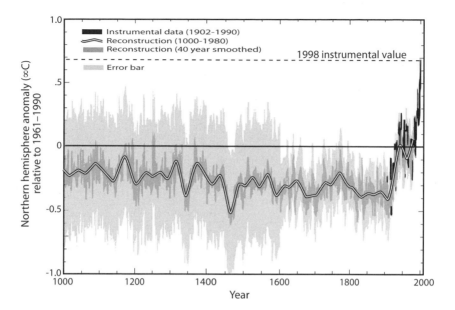

Fig. 10.5. The record of the last 1,000 years of temperature change (after Mann et al. 1999), which was stable until the late 1800s, when it suddenly shot upward in response to greenhouse gases released by the Industrial Revolution. The long straight line with the sudden kick upward gave it the nickname the "hockey stick curve." (Modified from image at Globalwarmingart.com; redrawn by Pat Linse)

net change before 1800, followed by the rapid curve upward at the end of the hockey stick.

Answering the Climate-Change Denialists

The data presented so far are clear-cut and beyond dispute among climate scientists. They represent an amazingly uniform consensus about climate change by nearly every scientist who works on the problem directly and publishes in top peer-reviewed journals. Such a group of scientists made up the IPCC, which was unequivocal in warning about the dangers of global warming and that humans were largely responsible. Their work was honored with the 2007 Nobel Peace Prize, shared with former vice president (and Oscar winner) Al Gore, who helped publicize it with his lectures depicted in the documentary film *An Inconvenient Truth*. The same basic conclusions have been endorsed by all leading scientific societies in nearly every nation of the world, and especially the most prestigious ones, the national academies of every nation that has spoken on the topic.

If that were not enough evidence, plenty of unbiased surveys show the depth and breadth of consensus among mainstream scientists that global warming is real and a serious threat to humanity. The most famous of these was conducted by Naomi Oreskes in 2004, which looked at all papers published on the topic in the world's leading scientific journal, *Science*, between 1993 and 2003. Of the hundreds of papers written by the world's top scientists, 980 supported global warming and none opposed it. A more recent study (Doran and Kendall Zimmerman 2009) found the support among the geophysical and atmospheric scientific community was virtually unanimous. This was confirmed when the prestigious *Proceedings of the National Academy of Sciences (USA)* published a paper showing that 98 percent of all scientists who actually publish climate research agree that global warming is real and caused by humans (Anderegg et al. 2010). I can think of no other topic in science (except evolution) that has this level of agreement among nearly all the world's leading scientists, especially among scientists who actually work with the scientific data and know the problem intimately. If it were not such a controversial topic politically, there would be almost no interest in discussing it among scientists, the evidence is so clear cut.

Right-wing politicians, Web sites, blogs, and Fox News continue to claim that global warming is still "controversial" or "unproven" and cite support in the scientific community. They can dredge up a few scientists here or there who are climate skeptics. One poll conducted by denialists led by Oklahoma senator James Inhofe (who receives major political contributions from the oil industry) attempted to show that there were at least a few hundred scientists who did not agree with the consensus about global warming. On closer inspection, the Center for Inquiry (CFI; www.centerforinquiry.net/) found that fewer than 10 percent of the names on the list had any appropriate credentials or direct experience in climate research. Instead, they were a mixture of scientists with no relevant training or experience. More than 80 percent had no refereed publications in climate science at all. About 4 percent of the denialists on the list actually protested their inclusion because they *favored* the IPCC 2007 consensus that global warming is real and man-made. Dr. Stuart Jordan, formerly a climate scientist for NASA and now with the CFI, wrote, "As a result of our assessment, Inhofe and other lawmakers using this report to block proposed legislation to address the harmful effects of climate change must face an inconvenient truth: while there are indeed some well respected scientists on the list, the vast majority are neither climate scientists, nor have they published in fields that bear directly on climate science" (Center for Inquiry 2007). Dr. Ronald Lindsey of CFI wrote, "Sen. Inhofe and others have had some success in conveying to the media the impression that the number of scientists skeptical about man-made global warming is swelling, yet this is demonstrably not true" (Center for Inquiry 2007). Inhofe's office had misleadingly claimed that the number of dissenting scientists was 13 times more than the number of United Nations scientists (52) who authored the 2007 IPCC. "But those 52 U.N. scientists were in fact summarizing for policymakers the work of over 2,000 active research scientists, all with substantially similar views on global warming and its causes. This is the kind of broadside against sound science and scientific integrity that we at CFI deplore" (Center for Inquiry 2007). The tactic of beating the bushes for any "expert" who will dissent from a legitimate scientific consensus resembles creationists' tactics. They frequently tout "lists of Ph.D. scientists who doubt evolution"; however, these lists consist of people with no relevant training, mostly nonscientists, plus a few engineers, chemists, or physicists. None of these "dissenting scientists" have firsthand research experience in geology, biology, or paleontology, and those who claim to be "geologists" or "biologists" got their degrees from creationist diploma mills. The National Center for Science Education cleverly parodied this misleading approach to cherry-picking names with "Project Steve" (http://ncseweb.org/taking-action/project-steve). It consists entirely of PhD scientists whose first name is Steve (less than 1% of all scientists) and who accept evolution—and *that list alone far outnumbers all the "creationist scientists" put together.*

Always check the credentials of people who write books about climate science. If they lack a PhD in climate science and are not actively researching climate science and publishing in respected journals of climate science, then they are amateurs in that topic and don't deserve to be taken seriously. This applies to many books and other writings that claim to show that there is no problem with global warming. For example, Bjorn Lomborg's book, *The Skeptical Environmentalist*, made a big splash when it first appeared, but Lomborg has no credentials in any science, and his work has been destroyed by many different groups of scientists who have revealed him to be incompetent in scientific issues (let alone his economic claims). More recently, people have carefully checked his footnotes and sources, and found that he has been quoting out of context (like a creationist), and most of his "sources" do not support the claims he makes in his book but have the opposite meaning. In August 2010, even Lomborg admitted that global warming was real and represented a serious threat to humanity.

Ian Plimer's recent book, *Heaven and Earth: Global Warming, the Missing Science* (2009) received more attention because Plimer is a geologist; however, Plimer is a mining geologist, not a climate scientist, and he (along with oil geologists) would be expected to have a vested interest in *not* understanding climate data that threatens his livelihood. His incompetence in climate science was revealed in numerous, scathing book reviews by both climate scientists and earth scientists (e.g., "Plimer's Homework Assignment," www.realclimate.org/index.php/archives/2009/08/plimers-homework-assignment/; "The Science is Missing from Ian Plimer's "Heaven and Earth," http://scienceblogs.com/deltoid/2009/04/the_science_is_missing_from_ia.php).

A favorite tactic of the anti-science movement such as creationism and climate denialism is to quote scientists out of context (called "quote mining") to support their position. When the full citation is consulted, it is clear that the misquotation contradicts the original author's intended meaning. This was particularly apparent in the attempt to distort the meaning in the wording of a few stolen e-mails from the Climate Research Unit of the University of East Anglia. If you read the actual e-mails and understand the context of the language used by climate scientists when talking casually to one another, it is clear that there was no great "conspiracy" or that they were faking data (www.realclimate.org/index.php/archives/2009/11/the-cru-hack/). This is one of the most dishonest tactics of all, because quote-miner climate denialists are either deliberately trying to mislead their audience by distorting the evidence or do not understand the quotes and their context in the first place.

Climate denialists, like creationists and Holocaust deniers, have many other traits in common. They pick on small disagreements between different labs as if scientists can't get their story straight, when in reality there is always a fair amount of give-and-take between competing labs as they try to get the answer right before the other lab does. When competing labs around the world have reached a consensus and arrive at the same answer, there is no longer any reason to doubt their common conclusion. The anti-scientists of creationism and climate denialism will also point to small errors by individuals to argue that the entire enterprise cannot be trusted. Scientists are human and can make mistakes, but the great power of the scientific method is that peer review weeds these out, so that when they speak with consensus, there is no doubt that their data are solid. Finally, the most convincing evidence that this is a purely political controversy, rather than a scientific debate, is that membership lists of creationists and climate denialists are mostly identical. Both are fed to overlapping audiences through right-wing media such as Fox News and Rush Limbaugh. The intelligent-design creationism Web site for the Discovery Institute (www.discovery.org/) lately posts items that have nothing to do with creationism at all but are focused on climate denialism and other right-wing causes.

A number of journalists and researchers have documented (McCright and Dunlap, 2003; Williams 2005; Curry et al. 2006; Mooney 2006, 2007; Hoggan, 2009; Oreskes and Conway, 2010) that the movement to deny global warming was not some rebellion scientists critical of the consensus spontaneously generated. It turns out to be a well-funded disinformation campaign supported by oil money to sow seeds of doubt about global warming. A number of memos and documents that were leaked out revealed that the entire movement was a giant public relations effort, not something real scientists would do in the unbiased pursuit of truth. Many have compared the denialist effort to how tobacco lobbyists tried to create a disinformation campaign of doubt. For decades, they forestalled any political action against tobacco by trying to prevent people from recognizing that smoking will kill you. In one particular instance, the oil lobby (through intermediaries) actually advertised and put up money to beat the bushes and find scientists—any scientists—who would dispute the consensus in some way. This is comparable to the practice of defense attorneys calling on highly paid, but unqualified, "expert" witnesses to confuse the jury about relatively clear-cut scientific facts. As Mooney (2006) and others documented, the Bush administration even had oil company lobbyists with no scientific qualifications placed high in their government bureaucracies. These unqualified political hacks edited government scientific reports to tone down the global warming emphasis or censored scientists altogether if their message was inconvenient to energy industry lobbyists.

Disinformation campaigns have had an effect. Numerous polls show that about 40 percent to 50 percent of Americans are still unsure about global warming or think that there is no scientific consensus. Numbers apparently improved after Al Gore's movie *An Inconvenient Truth* became a hit and changed the minds of many people, but still they represent a significant number of Americans who have bought into the disinformation campaign and don't trust the scientific consensus.

Let's set aside politics and look at the best available scientific data on why the climate science community considers global warming to be established beyond a reasonable doubt. Most common lies and myths about global warming are answered on the Web sites Grist (www.skepticalscience.com) and RealClimate (www.realclimate.org), so there is no need to repeat every argument here. I've already outlined the huge amount of climate data that have been available since 1999, and that data set keeps getting larger. Every new observation—from rapidly melting glaciers and ice caps to rising oceanic temperatures to rising sea level—is consistent with global climate change. The most common myth propagated by those who would deny global warming is that somehow this final 200 years of increasing temperature and carbon dioxide is part of a "natural cycle" of climate variability and nothing to be concerned about. As the "hockey stick curve" (fig. 10.5) shows, however, the rise in temperature is unprecedented over the past 1,000 years, and plots of the past 10,000 years show the same trend. The warming documented over the past century is faster and hotter than even the previous warmest period of the Holocene, or "Climatic Optimum," about 6,000–7,000 years ago.

Maybe 10,000 years is not convincing enough. Let's look at 680,000 years of climate change. We have such amazing records in long ice cores taken from Greenland and Antarctica. After these cores are drilled and brought up, they are carefully stored until trapped air bubbles from thousands to hundreds of thousands of years ago can be analyzed for their carbon dioxide and oxygen isotope contents. The most impressive of these is the EPICA core from Dome C on Dronning Maud Land, Antarctica (Siegenthaler et al. 2005; Spahni et al. 2005; fig. 10.6). It shows at least six or seven of the 100,000-year-long glacial-interglacial cycles, going back to almost 700,000 years ago. In the wiggly lines, you can see the carbon dioxide measurements for every stage of the glacials and interglacials, including the warmest periods of all seven previous interglacials. *At no time during any previous interglacial cycle did the carbon dioxide levels exceed 300 ppm, even at their warmest.* The earth's atmospheric carbon dioxide is already close to 400 ppm today and is headed to 600 ppm within a few decades, even if the release of greenhouse gases stopped immediately. This is decidedly *not* within the normal range of "climatic variability" but clearly unprecedented in human history. This is the point Gore made in *An Inconvenient Truth* when he stands atop the scissor lift and rides it upward to show the future levels of carbon dioxide on the long-term climate curve. It may have been a bit corny as a theatrical device, but it got the point across.

As Tim Flannery pointed out, ultimately the climate denialists are just a noisy distraction, a rearguard action. They are as irrelevant to the science of climate change as creationists are to the science of evolution. They may make a fuss and have friends in the Republican Party, but the reality of global warming has already convinced the most important audiences: world governments,

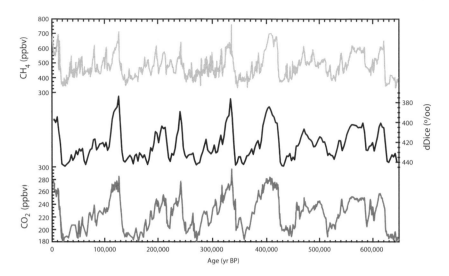

Fig. 10.6. Almost 800,000 years of temperature and climatic history (including at least seven interglacials) and carbon dioxide levels, as measured by the EPICA core from Antarctica (Modified from Siegenthaler et al. 2005, *Science* 310:1313–1317, Fig. 1. © American Association for the Advancement of Science. Used with permission.)

big businesses, the Pentagon, and other institutions that really *do* matter. The recent Copenhagen Climate Conference in December 2009 may not have produced as many results as some of us would have liked, or concrete pledges of how much each country would reduce emissions, with enforcement that some were hoping for. Nevertheless, it was a big step from the 1999 Kyoto conference because the United States, China, and India (three of the biggest players in global warming) all agreed to the seriousness of global warming and made some pledges to cut down their greenhouse emissions. A decade earlier, none of these countries agreed to the Kyoto pact.

If data are not convincing enough, and drowning polar bears aren't convincing, consider this: many normally conservative institutions (oil companies and many other businesses, the military, and insurance companies) are already planning on a world with global warming, even though they often will not admit it to right-wingers. Oil companies are already making inroads into other forms of energy, because oil will run out in a few decades or less and the effects of burning oil will make their business less popular. BP once stood for "British Petroleum," but now stands for "Beyond Petroleum." Oil companies may still spend a small percentage of their budget on alternative energy, but they are preparing for a future without oil. Even their publicity people know the public is aware of this problem and the need for alternative energy sources.

Pentagon sources indicate that the military has been making contingency plans for fighting wars in an era of global climate change: what will be the strategic threats when climate change alters the kinds of enemies we might fight and when water becomes scarce. The *New York Times* (2009) reported in "Climate Change Seen as Threat to U.S. Security" (www.nytimes.com/2009/08/09/science/earth/09climate.html?hp) that in December 2008 the National Defense University outlined plans for military strategy in a greenhouse world. The May 2004 issue of *Monthly Review* is full of articles about the Pentagon's plans for climate change. The issue was summarized and analyzed in a Pentagon report *An Abrupt Climate Change Scenario and Its Implications for United States National Security*, by Peter Schwartz and Doug Randall of the Global Business Network and commissioned by Peter Marshall, director of the Pentagon's Office of Net Assessment. Their report was released in October 2003 and laid out the grim scenarios that the military must consider in a greenhouse planet. The Pentagon report discusses the likelihood that agricultural decline and extreme weather conditions would overtax energy demand throughout the globe. Rich countries with resources, such as the United States and Australia, might build "defensive fortresses" to keep out hordes of immigrants, while the rest of the world fought over resources. As the report says, "Violence and disruption stemming from the stresses created by abrupt changes in the climate pose a different type of threat to national security than we are accustomed to today. Military confrontation may be triggered by a desperate need for natural resources such as energy, food and water rather than by conflicts over ideology, religion, or national honor. The shifting motivation for confrontation would alter which countries are most vulnerable and the existing warning signs for security threats" (p. 14). To the Pentagon, the big issue is global chaos and the potential for nuclear conflict. The world must "prepare for the inevitable effects of abrupt climate change—which will likely come [the only question is when] regardless of human activity" (*New York Times* 2009).

Insurance companies have no political axe to grind. If anything, they tend to be on the conservative side. They are simply in the business of assessing risk in a realistic fashion so they can accurately gauge their future insurance policies and what to charge for them. Yet they are all investing heavily in research on the disasters and risks posed by climatic change. In 2005, a study commissioned by the re-insurer Swiss Re said, "Climate change will significantly affect the health of humans and ecosystems and these impacts will have economic consequences" (Epstein and Mills 2005).

Right-wing ideologues may still not like the idea, but big businesses, including oil and insurance, and conservative institutions such as the military, don't have the luxury of living in an ideological ivory tower. They must plan for the world that will be in the next few decades; however, many symptoms are already apparent. They cannot afford to be caught flat-footed by climatic change when it threatens their survival. Neither can we as a society.

Brave New Greenhouse World

Many of global warming's effects on the planet are familiar, but let's get a few things straight. "Global warming" sounds like a good thing to the uninformed. A warmer world seems enticing when many of us shiver through cold winters, but it's much more complicated than that. Some places on earth will be warmer, but some will be cooler. The net effect is a few degrees warming over the entire planet, which is more than the total temperature change between glacial and interglacial stages during the ice ages. The most severe effects are already perceptible in the polar regions, where both the Arctic and Antarctic ice caps are melting at alarming rates (plate 15). Huge ice shelves the size of several U.S. states break away from Antarctica every few years, a phenomenon that has never been known to occur in human history. The Arctic ice cap is rapidly thinning and breaking up so that, in 2000, the Arctic Ocean at the North Pole was exposed to sunlight for the first time in almost 4 million years since the Arctic first froze over (see chapter 9). My former graduate advisor, the late Dr. Malcolm McKenna, was one of the first humans to see the North Pole covered in water and not ice (plate 15). Now this is an annual occurrence, and scientists are mapping the alarming disappearance of the Arctic ice sheet by satellite. More than 40 percent of the ice cover was lost between 1979 and 2003, and it is almost half gone as I write this. Already there is talk of ships taking the once-legendary "Northwest Passage" through the Canadian Arctic, because the region loses so much ice each year. This may give shippers a shorter route between the Atlantic and Pacific, but it is a serious problem for the rest of us, because an ice-free Arctic has a huge effect on the climate of the rest of the world.

The melting polar ice caps may be bad news for penguins, polar bears, Inuits, and Laplanders, but what about the human population of the world? It turns out that the effects in temperate latitudes are just as scary. More frequent heat waves and droughts, stronger and deadlier hurricanes, and all sorts of unexpected side effects have already occurred and will only worsen, because of global climate change. There are unexpected side effects as well. When the Rocky Mountains were colder with a sharp freeze each winter, the cold snap kept the pine bark beetle populations down. Now these beetles survive winters, and the pine forests of the Rocky Mountains have been devastated. Cold winters keep many diseases in check. Now a host of insect pests and pathogens are spreading new diseases. Cold-intolerant organisms are rapidly moving into higher latitudes because winters do not freeze as they used to. My wife grew up in southeastern Kansas, and at the time, there were no rattlesnakes or armadillos because of cold winters. Now both species are common in the region. This does not begin to address those species dying out because of climate change or how oceans (especially coral reefs) are dying off as they become too warm and too acidic because of excess carbon dioxide that seawater can no longer absorb.

The most serious effect of the last few episodes of greenhouse climate will be sea-level change. High sea levels of the geologic past drowned lower elevations of entire continents (see chapter 9). Currently, sea level is rising about 3–4 mm per year, more than 10 times the rate of 0.1–0.2 mm/year that has occurred over the past 3,000 years (fig. 10.7A). Before that, sea level was virtually unchanged over the past 10,000 years since the present interglacial began. This doesn't sound like much until you consider that the rate is accelerating and that most scientists predict it will rise 80–130 cm in the twenty-second century. A rise of 1.3 m (almost 4 feet) would drown many of the world's low-elevation cities, such as Venice, and many countries, such as Bangladesh. A number of tiny island nations such as Vanuatu and the Maldives, which barely stick above the ocean now, are already applying to the United Nations for relief, because they are about to vanish beneath the waves, and their entire population will have to move elsewhere (Environmental News Service, www .ens-newswire.com/ens/dec2005/2005-12-06-02.asp). If sea level rose by just 6 m (20 feet), nearly all the world's coastal plains and low-lying areas (the Louisiana bayous, Florida, most of the world's river deltas, the Netherlands, and the rest of Bangladesh) would be inundated. An overwhelming part of the world's population lives in coastal cities such as New York, Boston, Philadelphia, Baltimore, Washington, D.C., Miami, Shanghai, and London. All of those cities would be under water with just that much additional sea-level rise. If glacial ice caps on Antarctica and the Arctic Ocean melted completely (as they have several times before during greenhouse conditions), sea-level would rise by 65 m (215 feet)! The entire Mississippi Valley would flood, so you could dock your boat in Cairo, Illinois (fig. 10.7B). Such a sea-level rise would drown nearly every coastal region under hundreds of feet of water and inundate New York City, London, and Paris. All that would remain would be the tall landmarks, such as the Empire State Building, Big Ben, and the Eiffel Tower, which would be convenient for tying up boats but not much else.

There's one other terrifying dimension. As I discussed in chapter 9 of my book *Greenhouse of the Dinosaurs* (2009), this climatic change may not happen gradually. Studies of ice cores from the Younger Dryas episode, which marked the end of the last ice age and the beginning of the present Holocene interglacial about 11,000 years ago, showed that the switch from glacial to interglacial world did not happen over tens of thousands of years—but flipped wildly from one state to another in less than a decade! This is not the instantaneous

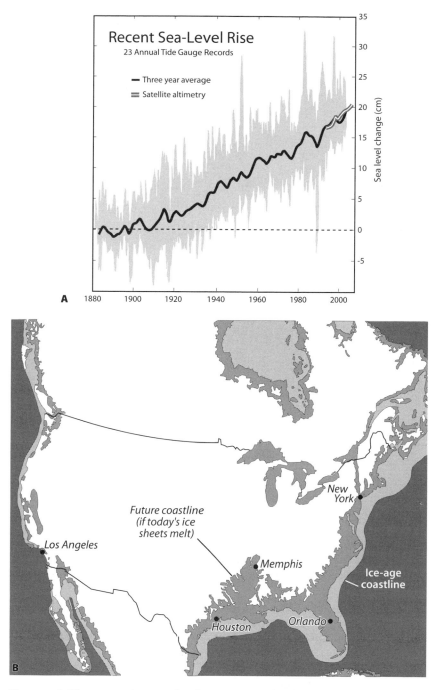

Fig. 10.7. A, The recent rise in sea level. (Courtesy Globalwarmingart.com; redrawn by Pat Linse) B. If all the glaciers melted, nearly all land below 215 feet in elevation would be drowned, and coastal plains, cities, and harbors would vanish. (From Prothero and Dott 2009; redrawn by Pat Linse)

event of the hokey Hollywood disaster movie *The Day after Tomorrow*, but it is much more rapid and catastrophic than anyone had ever thought before these ice cores were studied. Climate change is already more rapid than we have ever seen in the past, and apparently, it can happen even faster. As Richard Alley, an expert on ice-core climatology, wrote in *The Two-Mile Time Machine* (2000), the rapid switch from interglacial to glacial due to ocean currents "could produce a large event, perhaps almost as large as the Younger Dryas, dropping northern temperatures and spreading droughts far larger than the changes that have affected humans through recorded history, and perhaps speeding warming in the far south. The end of humanity? No. An uncomfortable time for humanity? Yes" (Alley 2000, 184).

During a speech on the Senate floor on July 28, 2003, Oklahoma Senator James Inhofe said, "With all of the hysteria, all of the fear, all of the phony science, could it be that man-made global warming is the greatest hoax ever perpetrated on the American people? It sure sounds like it." Inhofe needs to look at all the Cretaceous marine rocks in the western part of his state, or the Paleozoic marine rocks in the eastern and central part of his state, to realize that past greenhouse climates have drowned Oklahoma again and again. It would be ironic if someday future residents of Oklahoma, Texas, Arkansas, and Louisiana are sailing around their inundated prairie like Kevin Costner in *Waterworld*, wondering how someone as dogmatic and misinformed as Inhofe got elected and managed to delay necessary action on global warming for so long.

FOR FURTHER READING

Alley, R. 2000. *The Two-Mile Time Machine: Ice Cores, Abrupt Climate Change, and Our Future*. Princeton University Press, Princeton, NJ.

Archer, D. 2009. *The Long Thaw: How Humans Are Changing the Next 100,000 Years of Earth's Climate*. Princeton University Press, Princeton, NJ.

Broecker, W. S., and R. Kunzing. 2008. *Fixing Climate: What Past Climate Changes Reveal about the Current Threat—and How to Counter It*. Hill & Wang, New York.

Curry, J. A., P. J. Webster, and G. J. Holland. 2006. Mixing politics and science in testing the hypothesis that greenhouse warming is causing a global increase in hurricane intensity. *Bulletin of the American Meteorological Society* 87 (8): 1025–37.

Flannery, T. 2006. *The Weather Makers: How Man Is Changing the Climate and What It Means for Life on Earth.* Atlantic Monthly Press, New York.

Gore, A. 2006. *An Inconvenient Truth.* Rodale Press, Emmaus, PA.

Hoggan, J. 2009. *Climate Cover-Up: The Crusade to Deny Global Warming.* Greystone, Vancouver.

Linden, E. 2006. *The Winds of Change: Climate, Weather, and the Destruction of Civilizations.* Simon & Schuster, New York.

Mann, M. E., and L. R. Kump. 2008. *Dire Predictions: Understanding Global Warming.* DK Publishing, New York.

McCright, A. M., and R. E. Dunlap. 2003. Defeating Kyoto: the conservative movement's impact on U.S. climate change policy. *Social Problems* 50 (3): 348–73.

Mooney, C. 2006. *The Republican War on Science.* Basic Books, New York.

Mooney, C. 2007. *Storm World: Hurricanes, Politics, and the Battle over Global Warming.* Harcourt, New York.

Oreskes, N., and E. M. Conway. 2010. *Merchants of Doubt: How a Handful of Scientists Obscured the Truth on Issues from Tobacco Smoke to Global Warming.* Bloomsbury, New York.

Pearce, F. 2007. *With Speed and Violence: Why Scientists Fear Tipping Points in Climate Change.* Beacon Press, Boston.

Prothero, D. R. 2006. Our interglacial: The Holocene. Chap. 9, *After the Dinosaurs: The Age of Mammals.* Indiana University Press, Bloomington.

Prothero, D. R. 2009. Once and future greenhouse. Chap. 9, *Greenhouse of the Dinosaurs: Evolution, Extinction, and the Future of Our Planet.* Columbia University Press, New York.

Prothero, D. R., and R. H. Dott, Jr. 2009. *Evolution of the Earth.* 8th edition. McGraw-Hill, New York.

Siegenthaler, U., T. F. Stocker, E. Monnin, D. Lüthi, J. Schwander, B. Stauffer, D. Raynaud, J. M. Barnola, H. Fischer, V. Masson-Delmotte, and J. Jouzel. 2005. Stable carbon cycle–climate relationship during the Late Pleistocene. *Science* 310:1313–17.

Weiner, J. 1990. *The Next One Hundred Years: Shaping the Future of Our Living Earth.* Bantam Books, New York.

Williams, N. 2005. Heavyweight attack on climate-change denial. *Current Biology* 15 (4): R109–R110.

11

Mass Extinctions

When Life Nearly Died

Mass extinction is box office, a darling of the popular press, the subject of cover
stories and television documentaries, many books, even a rock song . . .
At the end of 1989, the Associated Press designated mass extinction as
one of the "Top 10 Scientific Advances of the Decade." Everybody has
weighed in, from the economist to National Geographic.
— David Raup, 1991

For every problem, there is a solution that is simple, neat and wrong.
— H. L. Mencken

Catastrophist Bandwagons

In 1980, a scientific paper hit the professions of geology and paleontology like
a blazing comet. After more than a century of speculating about how and why
the dinosaurs had vanished, and why other great mass extinctions occurred,
the authors had proposed a novel solution. In their scenario, about 65 million
years ago, an asteroid 10 km in diameter had slammed into the earth and
caused a global "nuclear winter" of cold and dark conditions that decimated

Fig. 11.1. The Gubbio boundary layer (dark band with the coin on it, just above the light-colored limestone). (Photo courtesy A. Montanari)

life. About 50 percent of the earth's species had vanished, including dinosaurs, marine reptiles, pterosaurs, and many marine organisms, especially among the plankton and the *Nautilus*-like ammonites. The paper provided a simple and neat solution to a complex problem, but this was not how the research started. Ironically, the authors were looking for something else altogether and found the evidence of an impact by accident.

In the late 1970s, a young geologist named Walter Alvarez was busy working on the geology of Italy, especially the Apennine Mountains that run down the spine of the peninsula. (I knew Walter when he was a postdoctoral student at Lamont-Doherty Geological Observatory of Columbia University and I was a graduate student.) Although Walter was primarily concerned with the way in which the mountain belt had grown and been deformed, he was aware that in some outcrops along a highway near Gubbio, Italy, there was an amazingly complete sequence of limestones that spanned the end of the Cretaceous Period (the final period of the Mesozoic or "age of dinosaurs") and the first few million years of the Tertiary Period ("the age of mammals"). Sandwiched between the limestones at this boundary was a thin layer of clay that represented the time when the extinction took place (fig. 11.1).

Walter took a sample of this Gubbio boundary rock home to his father,

Nobel Prize–winning physicist Luis Alvarez, at the University of California, Berkeley. They both were interested in finding a technique that would show how long it took the clay layer to form and, therefore, how rapid the mass extinction event had been. They looked for rare elements found primarily in extraterrestrial matter as a means of detecting how much cosmic dust had accumulated in the sample. If there was a lot of cosmic dust, the sample had accumulated slowly, but if there was very little, the time interval would be very short. They used the rare platinum-group element iridium as their tracer of cosmic dust and gave the samples to nuclear chemists Frank Asaro and Helen Michel to analyze. When the results came back, they were all stunned. The total amount of iridium (the iridium "anomaly") was much larger than any of them expected and could not be just the product of a simple rain of cosmic dust. Eventually, they concluded that it could have been produced by the impact of an extraterrestrial body, presumably an asteroid, which blasted a huge amount of crustal material into space upon landing and caused a "nuclear winter" of cold and darkness that killed off plants on land and algae in the ocean and so on up the food chain.

Naturally, the geological community immediately challenged such a bold and provocative hypothesis. All ideas in science must undergo the crucible of testing by other scientists and by peer review. At first, scientists thought it might be an artifact of the well-known and peculiar ability of clays to concentrate all sorts of rare materials or an artifact of unusual ocean chemistry, but those ideas were ruled out when the iridium anomaly was found in land sections. As the 1980s progressed, iridium anomalies were found at the boundary around the world, both in deep-sea cores and on marine sections that had been uplifted onto land. As a result, geologists and geochemists jumped on the bandwagon of declaring that it was the impact that killed the dinosaurs. Soon the media jumped into the fray. The May 6, 1985, issue of *Time* magazine ran an uncritical article and a crowd-pleasing cover of a *Tyrannosaurus rex* and a plummeting asteroid. The trendy impacts-extinction bandwagon was going full tilt, and many people jumped on for a free ride, funding for their research, and a chance to be part of the hottest idea in the profession.

Naturally, as soon as the impact advocates had argued that an asteroid caused the end-Cretaceous extinctions, other scientists quickly tried to find iridium anomalies and signs of impact at other great mass extinctions. Soon there were claims of iridium anomalies during the extinctions of the late Eocene 35.5 million years ago, the Devonian extinctions 375 million years ago, the late Triassic extinctions 200 million years ago, and the mother of all mass extinctions, the end of the Permian Period about 250 million years ago. By the late 1980s, many geologists were claiming that this "new catastrophism" of the importance of rare extreme events like extraterrestrial impacts was revolutionizing geology. At the International Geologic Congress in Washington, D.C., in 1989, the Canadian paleontologist Digby McLaren (who once suggested that a supernova caused the late Devonian extinctions) said that *all* mass extinctions were caused by impact, *whether or not there was evidence of impact in the fossil record!* I was in that audience and remember the stunned reaction of the audience at this untestable statement. The iconoclastic paleontologist David Raup (1991) wrote that all extinctions (even normal background extinctions) might be caused by impacts. With statements such as these, why bother gathering data any more? Impacts occurred, and extinctions occurred; therefore, all extinctions were caused by impacts. I remember receiving reviews by an anonymous scientist for an upcoming revision of my historical geology textbook *Evolution of the Earth* (written with Bob Dott). This reviewer said we should completely rewrite and reorganize the entire book to emphasize the importance of extraterrestrial impacts in earth history. I also recall having lunch with a major museum director at his swank downtown club to discuss setting up an institute devoted to the study of impacts and extinctions. He must be relieved that he didn't invest too much of his time and money in the idea, because it would have failed miserably based on advancements over the past 10 years.

The faddish impacts bandwagon in the 1980s was moving quickly, often far beyond (or despite) the data. As Keith Thomson (1988, 59) put it, "With most subjects there is a silly season, usually of unpredictable duration and with an intensity correlated with the status of the acceptance of the new idea, [including] proposal of ideas even more far out than the original one." Unfortunately, this describes much of the scientific media storm during the heyday of the impact bandwagon in the 1980s and 1990s. Claim after claim of iridium anomalies at this or that boundary, or some other cool-sounding discovery, were trumpeted in the press and spread through the scientific media. Reporters would parrot the provocative claim without any caveats or criticism, and seldom bothered to ask geologists and paleontologists who were not part of the bandwagon what they thought. Periodically, a reporter would call to ask me to comment on the latest news flash, but my criticisms would be reduced to one or two sentences or ignored completely. After all, the media want a simple, dynamic story, not one muddled by caveats and doubts. Yet even as these ideas were trumpeted in the media, other scientists were dropping their important research to test the ideas, scrutinize the original data, and see whether a different lab could replicate the results. Sure enough, a year or so later, the original "discovery" had been discredited, but the media took no notice. Nobody likes a killjoy. The media only want to feature the next startling claim in science, not a rebuttal of a year-old idea that reveals that their reporting was inadequate and misleading and their enthusiasm was premature. And so it goes—the media write up a flashy idea, most people only remember the initial intense media coverage, and almost no one but the specialists in the field remember that the original idea was debunked a year or two later.

Reality Check

Even though people are fascinated about radical and flashy ideas, other scientists have to buckle down and test these ideas to see whether they meet the critical scrutiny that science requires. Science is not a popularity contest or an academic field driven by fads with no external reality to check them against, like deconstructionism and so many other ideas elsewhere in academia. It is true that scientists are human and have their own biases, and they may follow the current fad like sheep. Eventually, though, ideas must be subjected to critical experiments and abandoned if they fail, no matter how popular they were. As Thomas Henry Huxley (1893, 6:8) put it, this is "the great tragedy of Science—the slaying of a beautiful hypothesis by an ugly fact."

Such is the case with the impact-extinction bandwagon. End-Cretaceous extinctions were popular with geochemists, geophysicists, planetary geologists, and many other geoscientists, but never widely accepted among paleontologists who knew this extinction event best. Even before the 1980 impact paper, paleontologists were aware that the end-Cretaceous event was a complex story that defied a simple explanation. Once thousands of researchers were working on the problem after the 1980 paper, a much more detailed record of the last few million years of the Cretaceous was collected, studied, and provided with a lot of detail that was not easily explained by the impact model.

In particular, it turned out that the end of the Cretaceous was a complicated time, with many events that could have contributed to the mass extinction. In addition to the impact (which was real, although it took until 1990 to establish this and find the site of the Chicxulub impact crater near Yucatán), there were two other major events happening at the same time. There was a big drop in sea level that dried up the shallow seaways that once flooded the continents (fig. 10.1) and certainly reduced the area of habitat for marine life. Finally, the end of the Cretaceous was marked by one of the biggest flood basalt eruptions in earth history, the Deccan lavas of western India and Pakistan. These eruptions produced more than 10,000 km^3 (2,400 cubic miles) of lava flows, with individual flows as thick as 150 m (500 feet), totaling at least 2.4 km (1.5 miles) in thickness. Such huge eruptions of material from the mantle would have filled the atmosphere with thick clouds of ash that carried iridium, as well as gases such as carbon dioxide that would have changed atmospheric and ocean chemistry.

Thus, there are three possible killers: (1) asteroid impact, (2) volcanoes, and (3) sea-level retreat. If the asteroid impact were the only important culprit, then almost all groups of organisms should have been severely affected and the extinction horizon should have been a single sharply defined event with all victims disappearing at the same level in the strata. If falling sea level or the

Deccan volcanoes were the killers, then a gradual decline in certain selected groups over the last few million years of the Cretaceous would have been expected, with relatively few victims right at the boundary.

In the marine record, there were a few types of plankton that seemed to die out abruptly (the planktonic foraminifera and the algae known as coccolithophorids), but the rest of the microplankton (the algae known as diatoms, plus the radiolarians and dinoflagellates) showed no effect. Although the squid-like ammonites vanished (either abruptly or slowly is subject to debate), nearly the rest of the groups of marine mollusks underwent only a minor extinction. Some, such as the cone-shaped rudistid clams and the huge, flat inoceramid clams, were both gone long before the impact occurred. The same is true of nearly all the rest of the marine invertebrates, almost none of which show a major abrupt extinction at the end of the Cretaceous (Prothero 2006, chap. 2; Prothero 2009, chap. 5; MacLeod et al. 1997). Likewise, the marine reptiles appear to have been in decline through most of the Cretaceous, and there's no evidence that any of them were alive to see the impact occur.

Even more striking is the evidence from terrestrial life (Archibald 1996). Yes, the dinosaurs vanished (except for those that evolved into birds). Some say that they did so abruptly, although there are no dinosaur fossils within the last 3 m (10 feet) of rock below the iridium layer, and dinosaurs were already in a long-term decline anyway. There was some change in the land flora, and a big "spike" of fern spores at the iridium anomaly, suggesting unusual conditions during that time. But virtually no other group of terrestrial animals was affected by this supposed "nuclear winter" catastrophe. Huge crocodilians and pond turtles all breezed through with no effect, as did bony fish, insects, and birds. There are only slight changes in the mammal fauna from marsupial dominated to placental dominated.

Nor was there any effect on the amphibians. Interestingly, amphibians provide an important test of a faddish idea: impact acid rain. Impact enthusiasts postulated that a large amount of acid rain produced from the debris was scattered by the impact. If that scenario actually happened, neither frogs nor salamanders would be alive today, because they are extremely sensitive to the acidity of fresh waters in which they live, and they breathe through their porous skins. Today, amphibians are vanishing from many places because of the tiny amount of human-induced acid rain in their habitats. If there were really global clouds of darkness, there would be no tropical bees left today, because they cannot survive more than a few days without flowers and are sensitive to cold. Some impact advocates have tried ad hoc explanations to salvage these inconvenient truths, such as claiming that the survivors were all aquatic or burrowing and hid out during the hellish days after the impact. That doesn't work for many of them (especially crocodilians, amphibians, birds, and insects). The terrestrial record does not support the "hell-on-earth" scenario that impact proponents promote. Yes, a rock from space clobbered Yucatán, but its effects must not have been as big or catastrophic as long asserted.

You would never know this by reading the media accounts or popular books on the topic, because they are uniformly biased toward the simplistic, spectacular scenario of asteroid impact. The impact model is still supported among geochemists and planetary geologists, who consider it "case closed," but not so with the paleontologists, who know the fossil record best. Several books by top-notch paleontologists (Keller and Macleod 1996; Archibald 1996; Hallam and Wignall 1997; Dingus and Rowe 1998) dispute the importance of the impact. A distinguished panel of 22 British paleontological specialists in nearly every group of marine fossils (MacLeod et al. 1997) argued against the impact scenario causing marine extinctions. In 2004, a poll (Brysse 2004) was conducted of the membership of the Society of Vertebrate Paleontology. Of those surveyed, 72 percent felt that the extinctions were caused by gradual processes followed by an impact. Only 20 percent felt that the impact was the sole cause. The other 8 percent had no opinion or questioned whether it was a mass extinction at all. After 30 years of the Cretaceous impact scenario, there is no clear consensus about the cause of the extinction, despite media reports.

"The Mother of All Mass Extinctions"

Although most people care only about dinosaurs and have never heard about any other kinds of prehistoric life, the end-Cretaceous extinctions were not even the biggest mass extinction in earth history. That honor goes to the end-Permian extinction 250 million years ago, which has been nicknamed "The Mother of All Mass Extinctions" (in reference to Saddam Hussein threatening the United States with the "Mother of all Battles" if they invaded Iraq in 1991). By various estimates, as much as 95 percent of all marine species, and a comparable percentage of land animals, died out at this event. The extinction decimated nearly every group of marine animals that had dominated the Paleozoic seafloor since the late Cambrian, more than 500 million years ago. Trilobites vanished, as did all the corals known at the time, and the tremendously abundant fusulinids, a group of amoeba-like protozoans with shells shaped like rice grains. Most of the other dominant Paleozoic groups (the brachiopods, or "lamp shells"; the bryozoans, or "moss animals"; the crinoids, or "sea lilies"; as well as the early ammonoids and most other mollusks) were reduced to a few surviving lineages to repopulate the planet in the Mesozoic. It was such a profound extinction that the entire composition of marine life was rearranged from a typical "Paleozoic fauna" to what is known as the "Modern fauna," which has ruled the oceans for the past 250 million years. Most land animals suffered severely, especially the once-dominant lineages of synapsids (once misleadingly called "mammal-like reptiles," although they are not really reptiles). These animals were pruned down to a few survivors, some of which evolved into the first true mammals by 200 million years ago.

Naturally, a mass extinction event as big as this has generated many possible explanations. During the 1970s and 1980s, a number of novel ideas were proposed. For example, the diversity of life is strongly controlled by the amount of habitable area for animals to live on. Geologists pointed out that the shallow marine shelf area was greatly reduced in the Permian because all the continents had coalesced to form Pangea. Unfortunately for this idea, it turns out that Pangea had already assembled by the early Permian, too early to have anything to do with the mass extinction. Likewise, geologists pointed to the great Permian ice cap (fig. 9.8A) and thought that the extinction might be due to global cooling. This idea was shot down because the ice sheet was already large by the Carboniferous Period and actually disappearing at the end of the

Permian. In fact, all the recent data showed that the end of the Permian was marked by a "super-greenhouse" global warming event and not by cooling after all.

In the 1990s, research intensified, and soon there were excellent sections of rock (especially in China) that yielded highly detailed records of the end of the Permian. When the time ranges of fossils were plotted, the Permian extinction suddenly looked more abrupt than had ever been expected. This led to a revival of the impact hypothesis, which had been proposed for the Permian event in the 1980s and then abandoned. Luann Becker and her colleagues (2001) have repeatedly pushed the idea that a large impact that produced Bedout Crater in Australia (Becker et al. 2004) was the culprit, and naturally, the press carried the story for months. Unfortunately, this exciting idea falls apart when the data are examined. No other laboratory can replicate Becker's claims of impact-derived chemicals, such as iridium, unusual helium, and "buckyballs" (the 60-carbon molecules known as buckminsterfullerenes) at the Permian-Triassic boundary (Farley and Mukhopadhyay 2001; Braun et al. 2001). To top it off, Bedout Crater is the wrong size and age to have anything to do with the end of the Permian (Glikson 2004; Koeberl et al. 2002; Renne et al. 2004; Wignall et al. 2004).

If not impacts, what could be the cause of "the Mother of All Mass Extinctions"? Currently, several good candidates have survived the gauntlet of testing and peer review. The most important is that some of the largest eruptions in earth history, the Siberian lavas, were spewing gigantic amounts of lava and greenhouse gases in a true "supervolcano" eruption at the end of the Permian (see chapter 3). These stacks of lava flows are up to 6.5 km (4 miles) in total thickness in 11 discrete eruptive sequences and cover a total of about 7 million square kilometers (2.7 million square miles), an area equivalent to the size of the continental United States (Erwin 2006). The ages of these flows have been recently redated between 252.2 million and 251.1 million years old, the same as the dates from the Permian-Triassic boundary in China. As mentioned earlier, geochemical evidence exists of runaway global warming, possibly triggered by carbon dioxide released from the Siberian eruptions. In addition, geochemical evidence from marine sediments suggests that the oceans were highly over-saturated in carbon dioxide (*hypercapnia*) and depleted in oxygen, a condition that would have been fatal to most kinds of marine organisms.

After the end-Permian and end-Cretaceous mass extinctions, the third biggest mass extinction occurred during the last two stages of the late Devonian, about 375 million years ago, when 75 percent of the marine species died out. Impacts were blamed at first, and iridium anomalies were reported. However, closer scrutiny shows that these iridium anomalies are at the wrong time, and the evidence for impacts (if it is real) is not correlated with the several pulses of geochemical changes and extinctions in the late Devonian (McGhee 1996).

The same can be said of the fifth largest mass extinction in earth history at the end of the Triassic Period about 200 million years ago. This event caused a significant extinction in marine life (especially certain types of brachiopods and most of the Triassic ammonoid lineages) and helped change the land fauna by wiping out archaic groups of reptiles and amphibians, replacing them with the newly evolved dinosaurs. Naturally, if geologists could tie the end-Cretaceous extinctions to impacts, they would try to do so with the end-Triassic extinctions that helped the dinosaurs take over the planet. Several geologists pointed to the huge Manicouagan Crater (fig. 11.2) in Quebec as the culprit (Olsen et al. 1987, 2002). This monstrous hole, which shows up as a huge ring on the satellite images of Quebec, is about 100 km in diameter, not much smaller than the Chicxulub crater in Yucatán associated with the Cretaceous impact. Unfortunately, recent dating of the crater at 214 million years ago is not even close to the age of the Triassic-Jurassic boundary at 201 million years, or near the age of any other mass extinction (Palfy et al. 2000). Shocked quartz and iridium have also been claimed for this boundary, but further scrutiny has shown that they are in such small concentrations that they were unlikely to be related to the extinction (Hallam 1990, 2004; Hallam and Wignall 1997; Tanner et al. 2004). Most recent research has focused on the huge eruptions of the Central Atlantic Magmatic Province (CAMP) basalts, which occurred when the Atlantic was ripping apart as Pangea broke up. However, many paleontologists do not regard the end-Triassic as a real "mass extinction

Fig. 11.2. The Manicouagan Crater in Quebec, photographed from the space shuttle. The impact that formed it was once blamed for the end-Triassic extinction, but it is now known to be an impact from a different time period. (Courtesy NASA)

event" but simply as an artifact of the compilation of the data (Tanner et al. 2004). If you plot all the true time ranges of organisms at very coarse resolution (as if they died only at the boundaries and not within the interval), you will get artificially high apparent "extinction rates" at the boundaries.

The most recent proposition that impacts cause extinctions was the claim by Firestone et al. (2007) that the extinction of ice age "megamammals" (large mammals more than 40 kg in weight) was due to the impact of an extraterrestrial object about 12,900 years ago. At first, the media had a field day with this proposition, and almost no dissenters or critics were heard from at all. Some geology textbooks even inserted this untested idea into their new editions without waiting for confirmation. As with other half-baked ideas from impact advocates, the "late Pleistocene impact" scenario has been discredited by a range of observations.

The late Pleistocene impact hypothesis was born from observations that there was a distinctive "black mat" organic layer in several localities across the southwestern United States, immediately above the last appearance of some ice age megamammal fossils. The victims of this megamammal extinction include not only huge mammoths and mastodons but also ground sloths, horses, camels, two genera of peccaries, giant beavers, and predators, such as short-faced bears, dire wolves, and saber-toothed cats but *not* bison, deer, pronghorns, and a number of other large mammals still found in North America today. The "black mat" is also above the first known artifacts of the Clovis culture, which were thought to be the first human arrivals from Eurasia and allegedly responsible for overhunting the megamammals to extinction. Firestone et al. (2007) also claimed to have found "nanodiamonds," iridium, helium-3, buckyballs, and many other geochemical and mineralogical "impact indicators" in the black mat layer and then painted a variety of different (and conflicting) scenarios about the impacting object (they are not consistent about whether it is a comet or an asteroid) hitting near the Carolina Bays region. This supposedly affected the Laurentide ice sheet in the northeastern part of North America and triggered the Younger Dryas cooling event at 12,900 years ago.

The entire scenario has been completely demolished by a number of lines of evidence. As Pinter and Ishman (2008) showed, there is no evidence of an impact in the Carolina Bays, and most of the alleged "impact evidence" is questionable when analyzed by other labs. Firestone et al. (2007) argued that the impact was an airburst, since there is no crater, no tektites, no shocked quartz, or no other high-pressure minerals, which are the best indicators of a true impact. Most of the material that was allegedly impact derived (nanodiamonds, iridium, helium-3, buckyballs, and so on) has been discredited by further testing or is also consistent with the normal rain of micrometeorites and not abundant enough to be evidence of an impact.

The claim that the black mat was an impact layer has been debunked. It likely indicates a high water table and wetter conditions associated with the abrupt Younger Dryas cooling event (Haynes 2008). The supposed "instantaneous" extinction of megamammals at this horizon has also been debunked because extinctions were scattered across a wide geographic area with different genera vanishing locally at different times (Grayson and Meltzer 2003; Fiedel 2009; Scott 2010). That mammoths, mastodons, giant deer ("Irish elk"), ground sloths, and many other megamammals did not die out at 12,900 years ago but survived in most cases to 10,000 to 11,000 years ago, discredits the idea that a single impact killed them off. In fact, *none* of the well-dated extinctions occur at 12,900 years ago. Most extinctions are either significantly younger than that interval (examples given earlier), or there are no good final dates for their last appearance. Little appears to have happened to the megamammals at precisely 12,900 years ago.

Particularly striking is the persistence of mammoths and ground sloths well into the Holocene (only 6,000 years ago), and, of course, the bison, deer, grizzly bear, cougars, peccaries, and pronghorns still with us; while elk and moose (which were not wiped out) came to North America at this time. The impact hypothesis does not explain the selectivity of this extinction. In addition, the South American, Australian, and Eurasian-African megafaunal extinctions are not synchronous with the alleged "impact," so it does nothing to explain their demise.

The claim that the "impact" had a severe effect on human cultures has been completely shot down as well (Buchanan et al. 2008), because there is no evidence that human cultures changed dramatically at this time or that there was a major population decline. Clovis culture was gradually transformed into Folsom, Dalton, and eastern U.S. Paleoindian cultures, and they apparently spread widely at this time, rather than declining. In early 2010, Jacquelyn Gill of the University of Wisconsin presented a paper at the Ecological Society of America meeting analyzing the details of lake sediments from the Northeast, which preserve a high-fidelity record of that time. She found no evidence of the impact debris that was supposed to be common, and her data were gathered even closer to the alleged impact site than the evidence garnered from the western United States. Nor was there any great shift in vegetation, pollen, spores, or other biotic signal that would be consistent with the impact hypothesis.

Finally, if the authors of the Pleistocene impact scenario had paid any attention to the past decade of research on impacts and extinctions (as discussed in this chapter), they would have realized that the "impacts cause extinction" notion is passé. As we have already shown, none of the great extinctions of the past (except possibly the end-Cretaceous event) are associated with impacts. In the next section, we will see that the largest impacts to hit the earth, other than Chicxulub, caused no extinction. It feels like the Firestone et al. (2007) impact scenario is a bad rehash of the debates from the 1980s. Apparently, the authors are still stuck on a bandwagon that has long since ground to a halt, except in the popular media. As Barnosky et al. (2004) and Scott (2010) showed, the causes of the late Pleistocene megafaunal extinctions are complicated and probably involve both human overhunting and climatic change. Impact, however, doesn't seem to be relevant.

The Fly in the Ointment

The perfect counterexample for all these trendy mass extinction scenarios comes from the interval of time I have studied for the past 30 years. Although these extinctions were not part of the "Big Five" mass extinctions, extinctions at the end of the Eocene and early Oligocene were significant, especially because so many archaic groups of the Eocene greenhouse world (see chapters 9 and 10) died out.

When the Alvarez et al. (1980) paper was first published and blamed the extinction of the dinosaurs on an impact, there was a rush to find evidence of iridium at other extinction horizons. Sure enough, they found evidence of iridium in the late Eocene (Asaro et al. 1982; Alvarez et al. 1982; Ganapathy 1982; Glass et al. 1982). They crowed that the discovery of iridium "solved" the mystery of the Eocene extinctions, and then they moved on to other research, believing the issue was resolved.

Those of us who have spent long years in the trenches, working on the details of the Eocene-Oligocene transition knew the story was not quite so simple. First, the iridium anomalies occurred in the *middle* of the late Eocene, 35.5–36.0 million years ago, but they are too young for the major extinction event at the end of the middle Eocene (37 million years ago) and too old for the early Oligocene glaciation event 33 million years ago. Second, there were almost no species known to die out at the time of the iridium anomaly (except for a handful of species of plankton known as radiolarians), and more than 99 percent of all life on the planet appeared to be unaffected by the impact event. The impact advocates had found iridium *close* to the Eocene-Oligocene boundary, but close doesn't cut it. Close only works in horseshoes and hand grenades. Thousands of feet of strata full of fossils lie between the 37-million-year-old extinction and the impact horizon, and thousands of feet more above the impact horizon and below the 33 million year event. This kind of slapdash, sloppy science is unacceptable and should never have been published in such prestigious scientific journals, except that every article on impacts was accepted without any serious criticism during the heyday of the impact bandwagon.

The story became more interesting when impact debris of the same age were found in oceanic cores, and finally when the craters themselves were identified (fig. 11.3). One of them is a huge buried crater responsible for the Chesapeake Bay (Poag 1999), and there is a slightly smaller one on the Atlantic continental shelf near Toms Canyon. The third crater, known as Popigai, was found in northeastern Siberia. These craters were truly impressive in size. The Chesapeake crater is almost 100 km (62 miles) in diameter, and the Popigai crater is about the same size; both are just slightly smaller than the 180 km (112 mile) diameter Chicxulub crater in Yucatán associated with the end-Cretaceous extinction. As the study of the Chesapeake crater continued, impact advocates were embarrassed to learn that such a large impact had almost zero effect on life in the late Eocene. Some had predicted that impact should

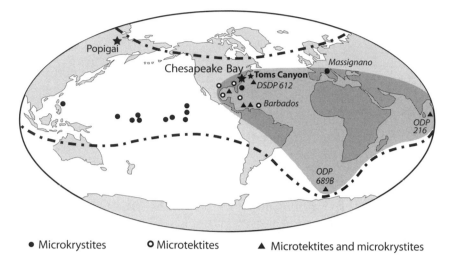

• Microkrystites   O Microtektites   ▲ Microtektites and microkrystites

Fig. 11.3. The location of the late Eocene impact craters and their debris fields. (After Poag 1997, *Palaios*; used by permission)

produce a ring of dust and debris and another "nuclear winter" scenario, but the evidence showed that the earth responded with warming in the latest Eocene instead. As discussed in Prothero (2009), the evidence for what caused the late Eocene extinctions is complex, but impacts had nothing to do with them, even though extraterrestrial objects did strike the planet.

During the height of the impact bandwagon, geologists were predicting that every impact should cause an extinction, and David Raup (1991) drafted a "kill curve" that would predict what percentage of species would go extinct for an impact of a given size (fig. 11.4). Raup had originally fit a simple curve to one data point, the end-Cretaceous event (dashed line in fig. 11.4). But with the addition of the nonextinction from the Chesapeake and Popigai craters, the "curve" changes shape radically (solid line in fig. 11.4). In its current incarnation, the Popigai-Chesapeake data in the Raup "kill curve" say that no extinction can be expected unless the impact is the size of Chicxulub or larger. This is borne out by studies that document thousands of impacts through earth history (Prothero 2004, 2007), but so far, only one—Chicxulub—is even potentially associated with a mass extinction. You can look up these thou-

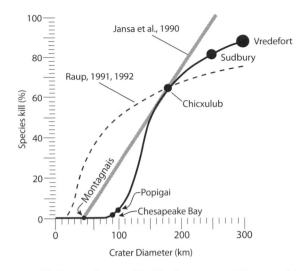

Fig. 11.4. The Raup "kill curve" as modified by Poag. (1997, *Palaios*; used by permission)

sands of impacts, their sizes, and estimated ages, on the online impact database (www.unb.ca/passc/ImpactDatabase/). As Prothero (2004) and Alroy (2002) both showed, there is no statistical correlation between any impacts or even the size of impacts, and the frequency of extinction.

The impact bandwagon seems to have run out of steam. In its heyday during the 1980s, planetary geologists received lots of publicity by claiming that asteroids were always threatening the earth and that we needed to spend millions on detecting them and thinking up ways to destroy or deflect any that might approach the earth. Big-budget Hollywood disaster films like *Deep Impact* and *Armageddon* made millions exaggerating this scenario. I do not question the basic idea that there are many rocks in space that could hit us, but what the geologic record shows is that *virtually no impact causes mass extinction*. Even impacts that hit during prehistoric times (including Meteorite Crater in Arizona, which hit 40,000 years ago) had no apparent effect on life except possibly in the area around the impact (Prothero 2004, 2007). It now seems bizarre and anachronistic to hear planetary geologists repeatedly raising this false alarm about the danger of impacts to drum up funding for their asteroid searches.

As Ward (2007) and many other former impact advocates now admit, the asteroid bandwagon is now passé, and geologists no longer maintain that impacts have an effect on life at all. They are now focusing on important changes in the atmosphere, especially "super-greenhouse" climates of high carbon dioxide and low oxygen, as more likely culprits for the end-Permian, end-Triassic, and several other major extinctions. This explanation (like the impact scenario and the massive volcanism scenario) fails for the late Eocene-Oligocene extinctions. That event is a perpetual "fly in the ointment" of every grandiose attempt to blame all mass extinctions on a simple single cause. It refuses to follow those simplistic scenarios that seduce some scientists.

The "Sixth Extinction"

In the early days of the impact-extinction frenzy, the late Jack Sepkoski of the University of Chicago compiled a huge database of the time range of all known families and genera of marine fossil. Dave Raup and Jack analyzed this data set with a number of computer techniques and found there was a general "background" rate of extinction (Raup and Sepkoski 1984). Five huge "mass extinction" events clearly stood above the background and were not just products of normal extinction intensities. We have already mentioned most of them. Number one was the end-Permian extinction 250 million years ago, when 95 percent of marine life vanished. The end-Cretaceous event that wiped out the dinosaurs and ammonites 65 million years ago was the second. The late Devonian event (375 million years ago), the late Ordovician event (450 million years ago), and the end-Triassic event (200 million years ago) rounded out the top five. The late Eocene extinctions were not among the "Big Five" events but came close to being the sixth largest extinction in earth history.

A mass extinction event is happening now, and threatens to dwarf even the biggest mass extinctions of the past. In *The Sixth Extinction*, Richard Leakey and Roger Lewin (1995) call this mass extinction event "the sixth extinction" to reflect that, over the past few centuries, humans are wiping out animals and plants at an unprecedented rate. Nobody knows exactly how many species have vanished and how fast extinction is occurring. No matter what estimates you read, the numbers are staggering. Biologists estimate that in the twentieth century between 20,000 and 2 million species went extinct. Current estimates place the rate of extinction at about 140,000 species per year. This is faster than any rate known in the geologic past.

Although we may not know the direct causes of any of the "Big Five" mass extinctions, there is no question about the culprit in the Sixth Extinction: *Homo sapiens*. Through either hunting or (more often) through habitat destruction and the introduction of diseases, pests, domesticated animals, and exotic invader species, humans have become the most dangerous and destructive species to inhabit the planet. No asteroid impact, gigantic volcanic eruption, or oversupply of carbon dioxide has ever been as destructive to life as humans have been. As Niles Eldredge (1991, 217) put it, "We are like loose cannons, able to wreak great damage on our own, and particularly dangerous if our effects happen to coincide with the physically induced changes that are also causing extinctions."

The single biggest factor in terrestrial extinction is our wholesale destruction of the world's rain forests. *Each day* 80 square miles of rain forest are lost, mostly in Africa, South America, and Southeast Asia. *Each year* an area of rain forest the size of Maine or Indiana vanishes. The cutting and burning of the rain forest not only destroys the habitats of more than half of the species on the planet but also contributes significantly to global warming. When the rain forest is cut down by slash-and-burn methods, the land is briefly turned into pastures for cattle to feed our fast-food obsession with hamburgers. Yet the rain forests' nutrients are locked in the trees, not the soil, so just a few years after the area is logged, it becomes a barren wasteland with poor soil that supports almost no life and quickly erodes to flood the rivers with sediment. Damage is now irreparable, and the rain forest cannot grow back. Meanwhile, poor subsistence farmers can no longer make a living on this wasteland and must move on and cut down more rain forest to survive.

People love to watch video footage of wild animals in nature, but most of the world's wild places are rapidly vanishing or have disappeared because of the invasion of humans and their domesticated animals. In his sad book *African Silences*, zoologist Peter Matthiessen (1991) surveyed tropical Africa and returned with grim news. Most of the great forests of West Africa were gone and what remains is a wasteland filled with starving people. What little remains of these rich forests (even within the supposed wildlife refuges), which once supported the bulk of Africa's wildlife, is now heavily poached by starving hunters looking for anything to eat as "bush meat." As Matthiessen put it, "The great silence that resounds from the wild land without a sign of human life, from which all the great animals are gone, is something ominous. Mile after mile, we stare down in disbelief" (52). The same has happened in the Congo Basin. We hear about the struggles to keep poachers from wiping out the last few populations of mountain gorillas studied by Dian Fossey, or chimpanzees studied by Jane Goodall, but they are just the tip of the iceberg. Most species in the African jungle are not cute, glamorous, or closely related to us, but they are already extinct or endangered. Beasts popularized in the movie *Born Free* and the TV show *Daktari* or from the explorations of Stanley and Livingstone are nearly gone, replaced by humans and their cattle. IMAX movies and documentaries on the Animal Planet channel have made us more conscious and sympathetic to the world of wild animals, but ironically, the real animals and their habitat have nearly vanished in the wild.

The plight of the oceans is a parallel tragedy. Marine biologists around the world have warned that the oceanic ecosystem is on the verge of collapse. Those who used to dive in the lush beauty of the tropical coral reefs are now reporting that most reefs are dead, destroyed by a combination of overfishing, pollution, and warming oceans, which kills off coral at an alarming rate. Nearly 75 percent of the world's fisheries are overexploited or at risk, and many fish (like the once-abundant Atlantic cod) are no longer fished because there are virtually none left. A 2003 study showed that 90 percent of the top predator species of the ocean have vanished, again largely due to overfishing. Sharks in particular display a catastrophic decline in numbers, because they reproduce slowly and do not recover quickly from excessive fishing. The 89 percent decline in hammerhead sharks is a typical number. Sharks are particularly vulnerable because they are accidentally caught in nets intended for other fish. The principal culprit has been the killing of sharks for their fins, since a single bowl of shark fin soup may sell for more than $100.

Some decline in marine species is due to disturbance of their habitat, especially where they breed and spawn, and the effects of warming and acidification of the oceans due to excess atmospheric carbon dioxide. The principal factor in the decline of fishing is the use of highly efficient, large-scale trawlers and fish-finding sonar, which can wipe out huge schools of fish in a single trawl. After this technology was introduced, catches nearly tripled from 1960 to 1992 and then plummeted when nearly all the major stocks of fish had been depleted.

Nothing encapsulates this story better than the plight of the bluefin tuna, which has long been considered a great delicacy among sushi aficionados. It has been so overfished that it is virtually extinct, and a single bluefin tuna can sell for as much as $173,000. Japanese fishing companies have already deep-frozen about 30,000 tons of bluefin, worth about $10 billion to $20 billion, and are holding on to it like any other precious commodity that will appreciate in value. Normally, it would be in our best interest to leave the fish alone and give its populations a chance to recover. Instead, its rarity makes it even more in demand, so it has become a status item like owning an expensive car or boat—if you're rich enough to afford bluefin sushi, you're at the top of the heap. Richard Ellis, author of *Tuna: A Love Story*, wrote the following:

People believe in their hearts that a piece of raw fish is worth $600. And one of the main reasons that it's worth $600 is because you can't afford it and I can't, but they can. That makes it very special, and it makes people who eat it special. Any kind of luxury goods largely come from that sort of statement: I can afford it, and you can't. I'll drive a Maserati, even if I can't drive it faster than 65 miles per hour in most of the United States. I can afford a $280,000 car, and you're stuck with a Dodge Neon. I can fly a private jet, drive a Maserati, do anything I bloody well please, including having a $600 piece of fish. And you can't. And this is the brutal truth: bluefin, which beyond their intrinsic value as living creatures happen to be one of the universe's more majestic species, a Platonic ideal of oceanic speed and grace, aren't being extinguished by our greed. They're being sacrificed to our vanity, pretension, and ostentation—the most pathetic of our vices. (Hive Mind 2009)

We are appalled by these stories of greed and stupidity or how overpopulation (especially the Third World) is driving most nondomesticated animals into extinction. Many people are highly motivated to do something to save the planet or to support efforts to reduce rain forest destruction or overexploitation of wild animals. Many people still react with, "Why should I care? Why should I worry about other species when humans are at risk too? After all, 99 percent of animals that have ever lived on this planet are extinct, so there's no denying the inevitable."

There are a number of answers to these questions, some practical and others moral and philosophical. Practical answers point out that wild nature provides huge benefits to us as humans. Many drugs come from rare tropical plants. Nature is essential to our food supply and for providing other needs—we could not survive without the wild kingdom. If we wipe out biodiversity before it has a chance to be studied, we miss the opportunity to identify and analyze tropical plants with the potential to provide valuable medicine, important natural pesticide, or other chemicals we require. The crash in the honeybee populations around the world threatens to destroy nearly all our agricultural crops because most depend on bees for pollination. As James Leape, the director general of the World Wildlife Federation, put it,

Reduced biodiversity means millions of people face a future where food supplies are more vulnerable to pests and disease and where water is in irregular or short supply. No one can escape the impact of biodiversity loss because reduced global diversity translates quite clearly into fewer new medicines, greater vulnerability to natural disasters and greater effects from global warming. The industrialised world needs to be supporting the global effort to achieve these targets, not just in their own territories where a lot of biodiversity has already been lost, but also globally. (Lovell 2008)

What is the big deal if we wipe out a few species here and there? Surely that won't cause the world to collapse or affect humans? In their 1981 book *Extinction*, Paul and Anne Ehrlich posed an interesting analogy. Suppose you were flying in a jetliner and looked out the window to see a rivet pop out of the wing. Then you saw the loss of another rivet, then another rivet, one after another. Perhaps one or two rivets will not damage the structural integrity of the aircraft and would not cause alarm. How many rivets are required before the wing falls apart and you crash? Would you be willing to perform this experiment and take the chance that you would die? That is comparable to the unintentional experiment we are performing on nature. Every species lost is another rivet that holds the world's ecosystems together. One or two species here or there may not make a big difference, but we really don't know how many can be lost before the entire planetary ecosystem collapses. With the dying coral reefs and the vanishing rain forest, we may already be beyond the turning point.

These are all practical reasons why we must strive to keep our fellow inhabitants of this small blue planet from vanishing. Many people regard conservation as an issue of philosophy, morality, and ethics. In their minds, humans suffer from too much hubris and anthropocentrism and treat the planet as though we were the only species that matters. They would argue that we have no more right to this planet than any other species, nor do we have the moral right to drive other species to extinction because we have the power to do so.

Finally, there are simple issues of esthetics as well. The earth is a beautiful place, and its many creatures are wonderful in their own right. It is tragic when we destroy them through our own carelessness, greed, apathy, and selfishness. The world without pandas, polar bears, and many other amazing creatures is truly a tragic hollow shell of its former self. In 1973, I traveled to Kenya, Tanzania, and Madagascar to see the amazing wildlife. Now, only 28 years later, nearly all that wildlife has been exterminated, and the African game parks have been depleted of rhinos and elephants, the favorites of poachers. My youngest son, Gabriel, is now 6 years old and is already a fanatic about animals, as I was at his age. He loves seeing them on TV and in his books, and he wants to go to the zoo as often as possible. I despair that he will ever get a chance to see most of the earth's great creatures in the wild because by the time he is old enough to see these creatures in their native habitat, most of them will be extinct in the wild.

In 1990, Mark Carwardine and the late Douglas Adams (author of the classic science fiction satire *Hitchhiker's Guide to the Galaxy*) published an amazing book, *Last Chance to See*. Adams and Carwardine visited many of the rarest animals on the planet before they vanished forever. In the process, they documented the efforts of a handful of dedicated wildlife biologists and conservationists as they risked their lives to save these rare creatures. As they put it in their book, "There is one last reason for caring, and I believe no other is necessary. It is certainly the reason why so many people have devoted their lives to protecting the likes of rhinos, parakeets, kakapos, and dolphins. And it is simply this: the world would be a poorer, darker, lonelier place without them" (Adams and Carwardine 1990, 211).

FOR FURTHER READING

Adams, D., and M. Carwardine. 1990. *Last Chance to See*. Harmony Books, New York.

Archibald, J. D. 1996. *Dinosaur Extinction and the End of an Era: What the Fossils Say*. Columbia University Press, New York.

Baskin, Y. 1998. *The Work of Nature: How the Diversity of Life Sustains Us*. Island Press, Washington, DC.

Ehrlich, P. R., and A. H. Ehrlich. 1981. *Extinction: The Causes and Consequences of the Disappearance of Species*. Random House, New York.

Ehrlich, P. R., and A. H. Ehrlich. 1998. *The Betrayal of Science and Reason: How Anti-Environmental Rhetoric Threatens Our Future*. Island Press, Washington, DC.

Eldredge, N. 1991. *The Miner's Canary*. Prentice-Hall, New York.

Erwin, D. 2006. *Extinction: How Life on Earth Nearly Ended 250 Million Years Ago*. Princeton University Press, Princeton, NJ.

Hallam, A., and P. B. Wignall. 1997. *Mass Extinctions and Their Aftermath*. Oxford University Press, Oxford.

Haynes, G., ed. 2009. *American Megafaunal Extinctions and the End of the Pleistocene*. Springer, New York.

Jansa, L. F., M.-B. Aubry, and F. M. Gladstein. 1990. Comets and extinctions: cause and effect? *Geological Society of America Special Paper* 247:223–232.

Kolm, K. A., and W. Reffalt. 1990. *Balancing on the Brink of Extinction: Endangered Species Act and Lessons for the Future*. Island Press, Washington, DC.

Leakey, R., and R. Lewin. 1995. *The Sixth Extinction: Patterns of Life and the Future of Humankind*. Doubleday, New York.

MacLeod, N., and G. Keller, eds. 1995. *Cretaceous-Tertiary Mass Extinctions: Biotic, and Environmental Changes*. W. W. Norton, New York.

Officer, C., and J. Page. 1996. *The Great Dinosaur Extinction Controversy*. Addison Wesley, New York.

Poag, C. W. 1997. Roadblocks on the kill curve: testing the Raup hypothesis. *Palaios* 12:582–90.

Poag, C. W. 1999. *Chesapeake Invader*. Princeton University Press, Princeton, NJ.

Prothero, D. R. 2004. Did impacts, volcanic eruptions, or climatic change affect mammalian evolution? *Palaeogeography, Palaeoclimatology, Palaeoecology* 214:283–94.

Prothero, D. R. 2006. *After the Dinosaurs: The Age of Mammals*. Indiana University Press, Bloomington.

Prothero, D. R. 2008. Do impacts really cause most mass extinctions? *In* J. Seckbach, ed., *From Fossils to Astrobiology*, 411–23. Springer, New York.

Prothero, D. R. 2009. *Greenhouse of the Dinosaurs*. Columbia University Press, New York.

Quammen, D. 1997. *The Song of the Dodo: Island Biogeography in an Age of Extinction*. Scribner, New York.

Raup, D. M. 1986. *The Nemesis Affair: A Story of the Death of the Dinosaurs and the Ways of Science*. W. W. Norton, New York.

Raup, D. M. 1991. *Extinction: Bad Genes or Bad Luck?* W. W. Norton, New York.

Raup, D. M. 1992. Large-body impacts and extinction in the Phanerozoic. *Paleobiology* 18:80–88.

Ward, P. 2007. *Under a Green Sky: Global Warming, the Mass Extinctions of the Past, and What They Can Tell Us about Our Future*. Smithsonian Books, Washington, DC.

12

Can We Survive Nature— and Our Own Folly?

I met a traveller from an antique land
Who said: Two vast and trunkless legs of stone
Stand in the desert. Near them on the sand,
Half sunk, a shatter'd visage lies, whose frown
And wrinkled lip and sneer of cold command
Tell that its sculptor well those passions read
Which yet survive, stamp'd on these lifeless things,
The hand that mock'd them and the heart that fed.
And on the pedestal these words appear:
"My name is Ozymandias, king of kings:
Look on my works, ye Mighty, and despair!"
Nothing beside remains. Round the decay
Of that colossal wreck, boundless and bare,
The lone and level sands stretch far away.
—Percy Bysshe Shelley, *Ozymandias*, 1818

A Sense of Proportion

We have come full circle and examined a range of natural disasters but now must step back and put them in perspective. Which ones cause the most dam-

Fig. 12.1. The toppled colossus of Ramses II (known as "Ozymandias" in earlier days) at the Ramasseum Temple in Luxor, Egypt, which inspired Shelley's poem. (Photo by S. F. E. Cameron, courtesy Wikimedia Commons)

age? Which disasters are the most dangerous to us in the short term? Which are dangerous in the long term? What is most likely to kill us? What is capable of not only hurting us as individuals but also destroying human civilization (fig. 12.1)?

Psychologists have shown that human beings are notoriously poor judges of relative risks and assessing which threats are serious and which are exaggerated. For deeply held psychological reasons, people are far more afraid of dying from a snakebite or in an earthquake, even though these are staggeringly improbable events for most U.S. citizens. Only 5 to 10 people die of snakebite each year, and as we mentioned in chapter 1, earthquakes kill, on average, six people per year. Yet because earthquakes are unpredictable and shatter our notion of "terra firma," we are unjustifiably afraid of them. Because snakes trigger a primordial fear response in our brain, we are terrified of them. When we were small, vulnerable hominids running across the African savanna, snakes were a threat because many African snakes, such as mambas and cobras, are poisonous. Now snakes are so heavily slaughtered in this country (even though most American snakes are nonpoisonous), we are more threatening to them than they are to us.

A more objective way of assessing real threats is to look at statistics, as an actuary or insurance adjuster does. An article by Borden and Cutter (2008) looked at deaths in the United States from all natural hazards from 1970 to 2004. Even though there were several sizable California earthquakes (1971, 1987, 1994) and large hurricanes during that time frame, they were not the number one killers. The top killers among natural hazards in this country are heat waves, storms, and winter (plate 16A).

Earthquakes, hurricanes, and other extreme weather events are terrifying disasters, but the biggest killers are slow and subtle: heat waves and drought. Likewise, we take severe storms and bitter cold winters for granted because they happen often, but they kill many more people than dramatic events such as tornadoes, hurricanes, and earthquakes. Hurricanes, earthquakes, and landslides are near the bottom of the list with fewer than 2 percent of the total deaths. Even though hurricanes and tornadoes are potentially dangerous, we are usually warned of their arrival, and most people take shelter or evacuate. Volcanic events did not make the list because the small Mount St. Helens eruption was the only deadly volcanic event in this country for better than a century.

Table 12.1. Estimated ranking of the top 10 deadliest natural disasters
in world history (excluding disease and famine)

Rank	Event	Location	Date	Death Toll
1	1931 China Floods	China	July–Nov. 1931	1,000,000–4,000,000
2	1887 Yellow River flood	China	Sept.–Oct. 1887	900,000–2,000,000
3	1556 Shaanxi earthquake	China	Jan. 23, 1556	830,000
4	1970 Bhola cyclone	Bangladesh	Nov. 13, 1970	500,000
5	1839 India cyclone	India	Nov. 25, 1839	300,000
6	526 Antioch earthquake	Byzantine Empire	May 20, 526	250,000
7	1976 Tangshan earthquake	China	July 28, 1976	242,000
8	1920 Haiyuan earthquake	China	Dec. 26, 1920	240,000
9	1975 Banqiao Dam Flood	China	Aug. 7, 1975	90,000–230,000
10	2004 Indian Ocean tsunami	Indian Ocean	Dec. 26, 2004	229,866

Source: List of natural disasters. Wikipedia. http://en.wikipedia.org/wiki/List_of_natural_disasters
_by_death_toll#Top_10_deadliest_natural_disasters_.5B1.5D

Borden and Cutter (2008) also plotted the risk on a county-by-county map of the United States (plate 16B). California is probably considered the most dangerous places to live in the United States because of its earthquakes, landslides, and brush fires; next would be the Gulf Coast with its deadly hurricanes. As the map shows, the opposite is true. Coastal California is one of the least hazardous locations (because it seldom has extreme killer weather of either the hot or cold variety). The Deep South is the most deadly region, where severe heat and humidity are common and occasional hurricanes occur. Also dangerous is "tornado alley" in the southern plains, with heat, drought, and tornadoes, and the southern Rocky Mountain region, with its desert heat and flash floods. The northern plains, the Rockies, and the Midwest are also death hot spots because of extreme cold, drought, and occasional flooding. The rest of the country does not show any striking trends one way or another. What you *don't* see is any strong correlation of high death risk with the fault zones map or even with the Gulf Coast–Florida hurricane zone.

Let's put that in a broader perspective. Many people are terrified of earthquakes, tornadoes, and hurricanes, but these events are uncommon except when warnings indicate that a hurricane or tornado is imminent. Heat waves and severe winter storms are more dangerous, but because we're accustomed to these events each year, we don't realize how deadly they can be. Worrying about natural disasters seems absurd compared with real risks that come from fast foods, driving, and smoking. Borden and Cutter (2008) point out that for the 20,000 people killed in the United States during the study period from 1970 to 2004, *there were 652,000 deaths from heart disease* (more than 30 times the natural disaster total). There were 600,000 deaths from cancer (also 30 times the total from natural disasters). Of cancer deaths, almost a third were from lung and other cancers due to smoking. Colorectal cancer, pancreatic cancer, prostate cancer, and breast cancer were the other biggest killers. There were 143,000 deaths from stroke, 130,000 from chronic lower respiratory diseases (bronchitis, pneumonia), and even 117,000 killed in accidents (mostly car accidents). If we took risk seriously and evaluated it objectively, we would worry about our diet and exercise, get frequent health checkups, stop smoking, and modify our driving habits. We may fear death in an earthquake or in a hurricane, but lunch, cigarettes, and driving are much more deadly!

A Global Perspective on Natural Disasters

The story changes significantly when we look at natural hazards worldwide. In the United States, building construction quality in seismic zones and our relatively good health care and emergency services reduce loss of life in earthquakes. In many underdeveloped countries such as Turkey, Iran, Armenia, Pakistan, Haiti, and China, and even European countries such as Greece and Italy, earthquakes have a much higher death toll, primarily because of the construction quality of buildings. Most building are constructed of simple stones or bricks held together by mortar, known as "unreinforced masonry" in the United States. Those buildings are death traps, even in a mild earthquake because they shake apart and then collapse with tons of weight on trapped inhabitants, killing thousands. By contrast, in seismically risky areas such as California, most older brick buildings have already shaken down in earlier earthquakes, and codes forbid any masonry except reinforced bricks, where steel tie rods and rebar is threaded through the holes in the cinder blocks to hold them together when buildings shake. For much of the Mediterranean–Alpine–southern Asia earthquake zone, however, the population is too poor to afford more expensive but seismically safer construction (and their governments do nothing to prevent substandard construction). Even as the survivors dig out their dead, they are rebuilding the same death traps with the same old bricks in the same way. In some cases, as with the 2008 Sichuan earthquakes in China, or the 1985 Mexico City earthquake (plate 3), buildings may be built of modern safer materials but still become death traps because of political corruption and shoddy construction.

The global list of the deadliest natural disasters (table 12.1) also includes more flood deaths than in the United States. As pointed out in chapter 5, most killer floods occurred in China. Their deadliness was exacerbated by political weakness or ineptitude in providing disaster relief. In addition, China's large population of peasant farmers living on floodplains have no place to go in a disaster. The same could be said for the effects of big typhoons, cyclones, and tsunamis in the Philippines, Bangladesh, China, and Burma. Events such as the recent Cyclone Nargis (chapter 6) in Burma were made far deadlier than necessary by the huge populations of poor peasants living on low-lying ground and vulnerable to big storms, and governments that can't or won't provide warnings, timely evacuation plans, or disaster relief in any large-scale meaningful way.

We must put these natural disasters in perspective. Hundreds of thousands of deaths in an earthquake, cyclone, or flood sound terrifying until we stack them up against the true killers: disease, famine, drought, and other slower but deadlier agents. If table 12.1 listed the mortality rates due to *all* deadly world events, all top ten events would be diseases and famines, and no natural disaster would make the top ten. The drought in India that led to the Great Famine of 1876–1878 killed at least 25 million people, at least 6 to 25 times as many as any natural disaster in Table 12.1. The 1918–1920 outbreak of Spanish influenza killed 20 million to as many as 100 million people, far more than any natural disaster in human history. The great bubonic plague, or "Black Death," outbreak of the 1300s may have killed as many as 75 million to 200 million, but the death estimates are uncertain.

Common diseases kill even more than the time-constrained "events" and "epidemics." Although the medical world is justifiably proud of containing it, smallpox killed at least 300 million in the twentieth century, and nobody knows how many humans have died of smallpox over the centuries. Measles killed more than 200 million over the past 150 years, even though it has been virtually eliminated in the United States. Malaria killed 80 million to 250 million people in the twentieth century, even though its spread and deadliness can be mitigated. Tuberculosis killed 40 million to 100 million people in the twentieth century, and its death rates are increasing as the world's skies become more polluted and the popularity of cigarettes increases outside the United States. Even AIDS, which is a relatively young epidemic (spreading only since the 1980s), has killed more than 25 million people worldwide, far more than any natural disaster.

Large and terrifying events such as tornadoes, landslides, and blizzards are impressive, but as we saw with U.S. statistics, they don't even rank. Worldwide, many of the more terrifying disasters in history are pikers when compared with death tolls such as those caused by diseases and famines. The world's deadliest tornado killed 1,300 in Bangladesh in 1989. The world's worst avalanche killed 96 in the United States in 1910. The world's deadliest blizzard killed 4,000 in Iran in 1972. The world's deadliest landslide killed 20,000 in Venezuela in 1999. The world's worst wildfire killed about 2,000 in the United States in 1871. The world's biggest historic volcanic eruption, Mt. Tambora in 1815, was a bit more impressive with 92,000 deaths, but still this doesn't come close to cracking the top ten list (table 12.1). The great heat wave that fried Europe in 2003 was the deadliest ever with more than 23,000 deaths, but that still does not make the top ten list. However much we are impressed by spectacular catastrophic events such as volcanoes, tornadoes, landslides, and blizzards, let's not mistake the terrifying power of these events with the death they cause. Disease and famine are a lot slower and less dramatic but far greater purveyors of death than any rapid catastrophic natural event.

A Choice of Catastrophes

This is the way the world ends
Not with a bang but a whimper.
—T. S. Eliot, *The Hollow Men*, 1925

By now, we've heard about how humanity can be killed off and civilization destroyed. We have also looked at events that seem to be catastrophic but turn out to be much less deadly than they seem. Most of these events have happened in recorded history or on human timescales, so they are familiar, even if we don't understand them well.

There's another category of agents of death and destruction due to forces largely outside the earth. Most often discussed by astronomers and science fiction fans, they operate on very long timescales, typically millions or billions of years. One of these agents, the impact of an asteroid or comet, is still popular in the media and among certain scientists. In chapter 11, we discussed the reality behind this, and what the latest data show about the actual risks of having civilization wiped out by the impact of an extraterrestrial object. Although an impact event would be a local disaster, the latest evidence shows that all but the biggest impacts have little or no effect on the earth as a whole and do not cause mass extinctions.

In 1979, the late great science author Isaac Asimov published an entire book on this topic, *A Choice of Catastrophes: The Disasters That Threaten Our World*. Ward and Brownlee (2000) discussed a variety of scenarios about how the earth might end. In 2009, Peter Ward published an anti-"Gaia hypothesis" book entitled *The Medea Hypothesis*, which also looks at long-term disasters on a million-year planetary scale. These books look at long-term risks from events far in the future. Among the commonly discussed scenarios are the following:

- *The sun becomes a red giant and incinerates the earth.* This scenario is pretty well understood by what we know about stellar behavior from the Hertzsprung-Russell diagram, which plots the fate of stars of different masses. Contrary to popular myth, the sun will not supernova, because it is not massive enough to do so. But it is massive enough that when its nuclear fuel begins to run out, it will expand into a red giant that will be 100 times larger than it is now and 2,000 times more luminous. In this scenario, the sun will flare out past the orbits of the inner planets, incinerating them all. The good news is that the timing for this scenario is well constrained based on looking at the histories of other stars in the universe. The best estimates put such an event approximately 5 billion years in the future, which is longer than the 4.5 billion years the earth has already existed.

- *The sun's increasing luminosity dries the earth up completely.* We might not have to wait for the sun to become a red giant because, as it reaches the end of its life, it will become much hotter and brighter with increased luminosity. (Conversely, 4 billion years ago it had a very low luminosity, and it was just barely bright enough to keep the earth from freezing.) Geologist James Kasting estimates that in only half a billion to a billion years from now, this event might occur, vaporizing the oceans and turning the earth into a barren desert, as are most terrestrial bodies in space (such as the Moon and Mars).

- *Some other star might become a supernova and destroy the earth with radiation.* In 2006, a supernova known as SN2006gy was discovered. It is one of the brightest supernovae ever observed, and it showed how common these bodies are across the universe. Fortunately, most are extremely far away, so their chances of sending anything lethal our way is small. For an object to have any effect on the earth, it must be closer than 100 light-years away. Only a handful of such objects are in space, so astronomers estimate that one of these might supernova every 20 million years on average. However, every search of the rocks representing more than a billion-year history of earth has turned up no reliable evidence that a supernova event struck our planet with enough radiation to leave a trace. Thus, even being generous, there may be another nearby supernova in less than 20 million years, but the event is still extremely unlikely to severely impact the earth.

Many other scenarios have been proposed: gamma-ray bursts, a nearby black hole, and even more far-fetched propositions, but calculations show that the probability that one of these events might threaten the earth is so small to not be worth considering (except as a plot device for science fiction). More realistically, these events are probably millions to billions of years in the future. Whether *Homo sapiens* will survive to experience them remains an open question.

There are quite a few scenarios based on reality, operating on short timescales (or are happening now), and worth considering. These include the following:

- *Global warming and the effects on climate and sea level.* As we discussed in chapter 10, global warming is already under way. It is predicted to cause all sorts of climatic and political chaos as storms, droughts, and rising sea level make parts of the earth inhospitable to people or destroy the ability to grow crops. The difficult question is whether it will make the earth uninhabitable and, if so, how soon.
- *The world's nuclear arsenals.* Even though the United States and the former Soviet Union have dismantled a small portion of the world's nuclear stockpiles since the end of the Cold War, thousands of nukes are still lying around in storage. Countries such as North Korea and Iran are desperately trying to get the bomb, and their leaders are unpredictable. The world has avoided mutually assured destruction and the "Dr. Strangelove scenario" using diplomacy and the fear that nuclear escalation could

destroy us all. That possibility is still real. What is different in the post–Cold War world is that rogue nations (North Korea and Iran) or even terrorist cells with nothing to lose might get a device, either from carelessness or through corruption. Experts are particularly concerned about the many nukes still lying around the former Soviet Union. The seriousness of this risk is difficult to assess, but the odds are certainly more likely than most of the scenarios of disaster from outer space. The scenarios of nuclear winter destroying life on this planet are realistic, given the size and number of weapons that still exist.

One overriding issue trumps every other disaster in its impact on our future as a species: *overpopulation.* As of February 2010, the official United Nations estimates place the world's population at 6.8 billion people. Since the end of the Black Death in the 1300s, global population has been increasing at a faster and faster rate, with recent rates at around 2 percent per year. For the first 4 million years of human existence, population was fairly stable or gradually increasing, held in check by war, disease, and famine. With the advent of modern medicine, which brought benefits of lowered infant mortality rates and increasing life spans, the "population bomb" has exploded. While there were 1.6 billion people alive in 1900, world population reached 3 billion by 1960, then 4 billion in 1974, only 14 years later. It will surpass 7 billion in a few more years. It will reach 9 billion to 11 billion by 2050, only four decades away. Fourteen to 15 babies are born every 6 seconds around the world. Each hour 8,800 babies are born, and each year at least 80 million more people are born.

Even though the attention on human population growth has been diverted elsewhere since the heyday of discussion in the 1970s, the problem has not gone away. Ultimately, it is at the root of most of the serious problems discussed in this book, from the "Sixth Extinction" and the destruction of wild lands, to the seriousness of natural disasters in crowded places such as China and Southeast Asia. Overpopulation influences how seriously the world suffers from famine and disease. Overpopulation drives the income inequality gap between rich and poor nations. Yet for a variety of political and economic reasons, overpopulation seems a hard issue to address. Dealing with overpopulation realistically is hampered when progrowth pundits say that the planet is just fine with 15 billion or 20 billion people, and certain religious authorities oppose family planning and drive poor, overpopulated countries into even greater poverty.

We assume that civilization is a permanent feature on this earth and that some form of our culture will persist indefinitely. As any archaeologist or historian knows, this is not true. We have many examples of extinct and vanished cultures that have left only a few durable artifacts, and we know little about how they lived or why they failed. One need only look at the mysterious Etruscans, Minoans, Mycenaeans, Mayans, or Anasazi. The list of failed societies is endless. Jared Diamond, in *Collapse: How Societies Choose to Fail or Succeed* (2004), points out that in instances in which we know why a society vanished, it is truly humbling. For example, the Easter Island culture vanished completely before the European settlers had much chance to witness their civilization, leaving only the famous huge stone heads, or *moai*, dotting the island (fig. 12.2). Diamond shows that, to a large extent, the extinction of the Easter Islanders was self-inflicted: too many people, too much overexploitation of their environment when they cut down all the trees on what had been a densely forested island, and finally starvation, disease, and warfare wiped out the survivors. As Diamond shows, such a fate could await our world civilization if we overpopulate this planet, damage our environment, or overexploit our resources. After all, 99 percent of all species that have ever lived are now extinct. There is no good biological reason to believe that our fate will be different, especially given our accelerated pace of self-destruction.

We are just temporary travelers on Spaceship *Earth*. Our genus *Homo* has been around less than 2 million years, only 1/2000 of the total history of the earth. *Homo sapiens* first appeared in modern form only about 100,000 years ago. The consensus of studies of the fossil record is that most species last only a few million years at most, and *Homo sapiens* exhibit all sorts of behaviors that suggest it will not survive much beyond the 100,000 years it has already existed. It's fun to speculate about the dangers of asteroid impacts, the sun burning up

Fig. 12.2. The famous giant stone heads, or *moai*, from Rano Raraku on Easter Island in the Pacific. As Jared Diamond has shown, the collapse of the Easter Island civilization is a classic example of how human societies can make bad choices (overpopulation, depletion of resources, destruction of their environment) and vanish, leaving only their monuments. (Photo by Artemio Urbina, courtesy Wikimedia Commons)

the earth as a red giant, and supernovae explosions, but much more prosaic fates are far likelier and more serious. If we can find ways to cope with climate change, to contain our nuclear arsenals, and to mitigate the effects of overpopulation, then we may well survive on this planet. Unfortunately, we act like bacteria, merrily reproducing and expanding in a Petri dish, oblivious to the threats that await us. Whether our intelligence and political will are good enough to save us from a dire fate is the challenge of the future.

FOR FURTHER READING

Asimov, I. 1979. *A Choice of Catastrophes: The Disasters That Threaten Our World.* Fawcett Columbine, New York.

Borden, K. A., and S. L. Cutter. 2008. Spatial patterns of natural hazards mortality in the United States. *International Journal of Health Geographics* 7:64.

Bostrom, N. 2002. Existential risks: analyzing human extinction scenarios and related hazards. *Journal of Evolution and Technology* 9. www.nickbostrom.com/existential/risks.html.

Chiarelli, B. 1998. Overpopulation and the threat of ecological disaster: the need for global bioethics. *Mankind Quarterly* 39 (2): 225–30.

Cutter, S. L. 2006. *Hazards, Vulnerability, and Environmental Justice.* Earthscan, London.

Diamond, J. 2004. *Collapse: How Societies Choose to Fail or Succeed.* Viking, New York.

Heinemann, W. 2003. *Our Final Century? Will the Human Race Survive the Twenty-First Century?* Arrow, New York.

Jensen, D. 2006. *Endgame.* Seven Stories Press, New York.

Rees, M. 2003. *Our Final Hour: A Scientist's Warning: How Terror, Error, and Environmental Disaster Threaten Humankind's Future in This Century—on Earth and Beyond.* Basic Books, New York.

United Nations. 2006. *World Population Prospects.* www.un.org/esa/population/publications/wpp2006/WPP2006_Highlights_rev.pdf.

United Nations Office for the Coordination of Humanitarian Affairs (OCHA). 2009. Climate change: humanitarian impact. http://ochaonline.un.org/News/InFocus/ClimateChangeHumanitarianImpact/tabid/5099/language/en-US/Default.aspx.

Ward, P. D. 2009. *The Medea Hypothesis: Is Life on Earth Ultimately Self-Destructive?* Princeton University Press, Princeton, NJ.

Ward, P. D., and D. Brownlee. 2000. *Rare Earth: Why Complex Life Is Uncommon in the Universe.* Springer, New York.

Ward, P. D., and D. Brownlee. 2004. *The Life and Death of Planet Earth: How the New Science of Astrobiology Charts the Ultimate Fate of Our World.* Holt, New York.

Wilson, E. O. 2002. *The Future of Life.* Knopf, New York.

Bibliography

Allen, J. E., M. Burns, and S. C. Sargent. 1986. *Cataclysms on the Columbia.* Timber Press, Portland, OR.

Alley, R. 2000. *The Two-Mile Time Machine: Ice Cores, Abrupt Climate Change, and Our Future.* Princeton University Press, Princeton, NJ.

Alroy, J. 2002. Extraterrestrial bolide impacts and biotic change in North American mammals. *Journal of Vertebrate Paleontology* 22 (3): 32A.

Alroy, J. 2003. Cenozoic bolide impacts and biotic change in North American mammals. *Astrobiology* 3 (1): 119–32.

Alvarez, L. W., W. Alvarez, F. Asaro, and H. V. Michel. 1980. Extraterrestrial cause for the Cretaceous-Tertiary extinction. *Science* 208:1095–1108.

Alvarez, W., F. Asaro, H. V. Michel, and L. W. Alvarez. 1982. Iridium anomaly approximately synchronous with terminal Eocene extinctions. *Science* 216:886–88.

Anderegg, W. R. L., J. W. Prall, J. Harold, and S. H. Schneider. 2010. Expert credibility on climate change. *Proceedings of the National Academy of Sciences (USA)* 107:12107–9.

Asaro, F., L. W. Alvarez, W. Alvarez, and H. V. Michel. 1982. Geochemical anomalies near the Eocene/Oligocene and Permian/Triassic boundaries. *Geological Society of America Special Paper* 190:517–28.

Barnosky, A. D., P. L. Koch, R. S. Feranec, S. L. Wing, and A. B. Shabel. 2004. Assessing the causes of late Pleistocene extinctions on the continents. *Science* 306:70–75.

Becker, L., R. J. Poreda, A. R. Basu, K. O. Pope, T. M. Harrison, C. Nicholson, and R. Iasky. 2004. Bedout: a possible end-Permian impact crater offshore of northwestern Australia. *Science Express Research Article* 10.1126/science.1093925.

Becker, L., R. J. Poreda, A. G. Hunt, T. E. Bunch, and M. Rampino. 2001. Impact event at the Permian-Triassic boundary: evidence from extraterrestrial noble gases in fullerenes. *Science* 291:1530–33.

Branner, J. C. 1913. Earthquakes and structural engineering. *Bulletin of the Seismological Association of America* 3 (1): 1–5.

Braun, T., E. Osawa, C. Detre, I. Toth. 2001. On some analytical aspects of the determination of fullerenes in samples from the Permian/Triassic boundary layers. *Chemical Physics Letters* 348:361–62.

Brysse, K. 2004. Off-limits to no one: vertebrate paleontologists and the Cretaceous-Tertiary mass extinction. PhD diss., University of Alberta, Alberta, Canada.

Buchanan, B., M. Collard, and K. Edinborough. 2008. Paleoindian demography and the extraterrestrial impact hypothesis. *Proceedings of the National Academy of Sciences (USA)* 105:11651–54.

Center for Inquiry. 2007. Ranking member's Senate Minority Report on global warming not credible, says Center for Inquiry. July 17. http://www.center forinquiry.net/opp/news/senate_minority_report_on_global_warming _not_credible/

Curry, J. A., P. J. Webster, and G. J. Holland. 2006. Mixing politics and science in testing the hypothesis that greenhouse warming is causing a global increase in hurricane intensity. *Bulletin of the American Meteorological Society* 87 (8): 1025–37.

Dingus, L., and T. Rowe. 1998. *The Mistaken Extinction: Dinosaur Evolution and the Origin of Birds*. W. H. Freeman, New York.

Doran, P., and M. Kendall Zimmerman. 2009. Examining the scientific consensus on climatic change. *EOS* 90 (3): 22.

Epstein, P. R., and E. Mills, eds. 2005. *Climate Change Futures: Health, Ecological and Economic Dimensions*. Center for Health and the Global Environment, Harvard Medical School, Boston, MA.

Farley, K. A., and S. Mukhopadhyay. 2001. An extraterrestrial impact at the Permian-Triassic boundary? Comment. *Science* 293:2343.

Fiedel, S. 2009. Sudden deaths: the chronology of terminal Pleistocene mega-faunal extinction. *In* G. Haynes, ed., *American Megafaunal Extinctions at the End of the Pleistocene*, 21–38. Springer, New York.

Firestone, R. B., et al. 2007. Evidence for an extraterrestrial impact 12,000 years ago that contributed to the megafaunal extinctions and the Younger Dryas cooling. *Proceedings of the National Academy of Sciences(USA)* 104:16016–21.

Ganapathy, R. 1982. Evidence for a major meteorite impact on the earth 34 million years ago: implications for Eocene extinctions. *Science* 216:885–86.

Glass, B. P., D. L. DuBois, and R. Ganapathy. 1982. Relationship between an iridium anomaly and the North American micro-tektite layer in core RC9-58 from the Caribbean Sea. *Journal of Geophysical Research* 87:425–28.

Glikson, A. Y. 2004. Comment on "Bedout: a possible end-Permian impact crater off northwestern Australia." *Science* 306:613.

Grayson, D. K., and D. J. Meltzer. 2003. A requiem for North American overkill. *Journal of Archeological Science* 30:585–93.

Green, N. C. 1900. *Story of the Galveston Flood.* R. H. Woodward Co., Baltimore.

Hallam, A. 1990. The end-Triassic mass extinction event. *Geological Society of America Special Paper* 585:577–83.

Hallam, A. 2004. *Catastrophes and Lesser Calamities: The Causes of Mass Extinctions.* Oxford University Press, Oxford.

Haynes, C. V., Jr. 2008. Younger Dryas "black mats" and the Rancholabrean termination in North America. *Proceedings of the National Academy of Sciences (USA)* 105:6520–25.

Hive Mind. 2009. Frank Bruni, rich a**holes and bluefin tuna as Gucci handbags. True/Slant blog post. http://trueslant.com/hivemind/2009/07/24/frank-bruni-rich-aholes-and-bluefin-tuna-as-gucci-handbag/.

Huxley, T. H. H. 1893. *Collected Essays,* vol. 6. Macmillan, London.

Keller, G. 2005. Impacts, volcanism, and mass extinction: random coincidence or cause and effect? *Australian Journal of Earth Sciences* 52:725–57.

Keller, G., and N. MacLeod, eds. 1996. *Cretaceous-Tertiary Mass Extinctions: Biotic and Environmental Changes.* W. W. Norton, New York.

Koeberl, C., K. A. Farley, B. Peucker-Ehrenbrink, and M. A. Sephton. 2002. Geochemistry of the end-Permian extinction event in Austria and Italy: no evidence for an extraterrestrial component. *Geology* 32:1053–56.

Lorenzo Dow's Journal. 1849. Published by Joshua Martin; printed by John B. Wolff.

Lovell, J. 2008. World species dying out like flies says WWF. Reuters, May 16. Environmental News Network. www.enn.com/wildlife/article36390.

MacLeod, N., et al. 1997. The Cretaceous-Tertiary biotic transition. *Journal of the Geological Society, London* 154:265–92.

Mann, M. E., R. S. Bradley, and M. K. Hughes. 1999. Northern Hemisphere temperatures during the past millennium: inferences, uncertainties, and limitations. *Geophysical Research Letters* 26:759–62.

Mann, M. E., J. D. Woodruff, J. P. Donnelly, and Z. Zhang. 2009. Atlantic hurricanes and climate over the past 1,500 years. *Nature* 460:880–83.

Matthiessen, P. 1991. *African Silences.* Random House, New York.

McGhee, G. R., Jr. 1996. *The Late Devonian Mass Extinction.* Columbia University Press, New York.

McGhee, G. R., Jr. 2001. The "multiple impacts hypothesis" for mass extinction: a comparison of the late Devonian and late Eocene. *Palaeogeography, Palaeoclimatology, Palaeoecology* 176:47–58.

Mooney, C. 2006. *The Republican War on Science*. Basic Books, New York.

Mooney, C. 2007. *Storm World: Hurricanes, Politics, and the Battle over Global Warming*. Harcourt, New York.

Morning Edition. Remembering the 1906 San Francisco earthquake. 2006. Transcript. Renee Montaigne, host. NPR News. http://www.npr.org/templates/transcript/transcript.php?storyId=5334411.

Officer, C., and J. Page. 1993. *Tales of the Earth*. Oxford University Press, New York.

Olsen, P. E., D. V. Kent, H.-D. Sues, C. Koeberl, H. Huber, A. Montanari, E. C. Rainforth, S. J. Fowell, M. J. Szajna, and B. W. Hartline. 2002. Ascent of dinosaurs linked to an iridium anomaly at the Triassic-Jurassic boundary. *Science* 296:1305–7.

Olsen, P. E., N. H. Shubin, and M. H. Ander. 1987. New early Jurassic tetrapod assemblages constrain Triassic-Jurassic tetrapod extinction event. *Science* 237:1025–29.

Oreskes, N. 2004. Beyond the ivory tower: the scientific consensus on climatic change. *Science* 306:1686.

Palfy, J., J. K. Mortensen, and E. S. Carter. 2000. Timing the end-Triassic mass extinction: first on land, then on sea? *Geology* 51:171–72.

Pellegrino, C. 1991. *Unearthing Atlantis*. Random House, New York.

Pinter, N., and S. E. Ishman. 2008. Impacts, mega-tsunami, and other extraordinary claims. *GSA Today* 18 (1): 37–38.

Pliny the Younger. 1963. *Letters of the Younger Pliny*. Trans. B. Radice. Penguin, New York.

Plummer, C. C., and D. H. Carlson. 1999. *Physical Geology*. 8th ed. McGraw-Hill, New York.

Prothero, D. R. 2007. *Evolution: What the Fossils Say and Why It Matters*. Columbia University Press, New York.

Raffles, S. 1830. *Memoir of the life and public services of Sir Thomas Stamford Raffles, F.R.S. &c., particularly in the government of Java 1811–1816, and of Bencoolen and its dependencies 1817–1824: with details of the commerce and resources of the eastern archipelago, and selections from his correspondence*. John Murray, London.

Raup, D. M., and J. J. Sepkoski, Jr. 1984. Periodicity of extinctions in the geologic past. *Proceedings of the National Academy of Sciences (USA)* 81:801–5.

Renne, P. R., H. J. Melosh, K. A. Farley, W. U. Reimold, C. Koeberl, M. R. Rampino, S. P. Kelly, and B. A. Ivanov. 2004. The Bedout Crater—no sign of impact. *Science* 306:610.

Roosevelt, T. 1888. *Ranch Life and the Hunting Trail*. New York, Century Co.

Scott, E. 2010. Extinctions, scenarios, and assumptions: Changes in latest Pleistocene large herbivore abundance and distribution in western North America. *Quaternary International* 217:225–39.

Spahni, R., J. Chappellaz, T. F. Stocker, L. Loulergue, G. Hausammann, K. Kawamura, J. Flückiger, J. Schwander, D. Raynaud, V. Masson-Delmotte, and J. Jouzel. 2005. Atmospheric methane and nitrous oxide of the Late Pleistocene from Antarctic ice cores. *Science* 310:1317–21.

Tanner, L. H., S. G. Lucas, and M. G. Chapman. 2004. Assessing the record and causes of Late Triassic extinctions. *Earth Science Reviews* 65:103–39.

Tappan, E. M., ed. 1914. *The World's Story: A History of the World in Story, Song, and Art*. 14 vols. Vol. 5, *Italy, France, Spain, and Portugal*. Houghton Mifflin, Boston.

Thomas, T. P. 1964. An Alaskan family's night of terror. *National Geographic* 126, no. 1 (July): 142–56.

Thomson, K. S. 1988. Anatomy of the extinction debate. *American Scientist* 76:59–61.

Twain, M. 1871. *Roughing It*. American Publishing Co., Hartford, CT.

Twain, M. 1883. *Life on the Mississippi*. Signet Classic Edition. Bantam Books, New York.

The West Film Project and WETA. 2001. Rain follows the plow. Episode 7, *The West*. PBS. www.pbs.org/weta/thewest/program/episodes/seven/rainfollows.htm.

Wignall, P. B., B. Thomas, R. Willink, and J. Watling. 2004. The Bedout crater—no sign of impact. *Science* 306:609.

Index

Page numbers in boldface denote illustrations.